'How the media communicates risk and uncertainty to their audiences is a critical issue. It is vitally important that the public are aware of the risks associated with unmanaged climate change. As this book points out, the media need to convey this information clearly and accurately without resorting to scaremongering. It highlights both good and bad practice by the media and provides extremely sensible suggestions for improvements in the future.'

Lord (Nicholas) Stern, Chair of the Grantham Research Institute on Climate Change and the Environment, London School of Economics and Political Science

'Our climate system is very complex, and predictions about its future are inevitably uncertain. However, uncertainty does not imply inaction. Rather, good decision making in climate-sensitive sectors will require not only predictions of the most likely climate change, but also reliable estimates of the uncertainty in future climate change. This report makes a valuable contribution in describing uncertainty prediction in climate science and highlights areas where this area of science could be promoted better, in the media and elsewhere.'

Tim Palmer, Royal Society Research Professor in Climate Physics, Oxford University

'Communicating the observed and potential consequences of climate change is a challenging task, one that is often done poorly in the media. This important book provides many valuable insights into the use of a risk framework to communicate climate change. It is essential reading for those in the climate change communication business, especially those in the media who want to take an informed and professional approach to the topic.'

Will Steffen, Climate Commissioner, Australia

'This is a really valuable book about the framing of the climate debate as an issue of 'risk', emphasising that, in spite of inevitable uncertainty about the future, we can still take appropriate action to hedge against bad outcomes. It is heartening to see the dual role of number and metaphor in communication – my favourite analogy is with setting up a good pension as a sensible precaution for the highly likely, but of course not certain, prospect of an extended old age.'

David Spiegelhalter, Winton Professor for the Public Understanding of Risk, Cambridge University

'This is essential reading for any scientist, including IPCC authors, who are in the business of communicating risk and uncertainty about climate change to policy makers, the public or the media.'

Cecilie Mauritzen, Director, Center for International Climate and Environmental Research (CICERO), Oslo Norway

'The latest in a series of incredibly useful and important studies shedding light on one of the most important, yet complicated and misunderstood, stories in journalism.'

Curtis Brainard, Contributing Editor, *Columbia Journalism Review*

'It's as hard to find the right language to convey climate change as it is to explain love or describe jazz. This abundantly referenced study is an essential guide for journalists through the new language of risk.'

Richard Black, former BBC environment correspondent

RISJ CHALLENGES

CHALLENGES present findings, analysis and recommendations from Oxford's Reuters Institute for the Study of Journalism. The Institute is dedicated to the rigorous, international comparative study of journalism, in all its forms and on all continents. CHALLENGES muster evidence and research to take forward an important argument, beyond the mere expression of opinions. Each text is carefully reviewed by an editorial committee, drawing where necessary on the advice of leading experts in the relevant fields. CHALLENGES remain, however, the work of authors writing in their individual capacities, not a collective expression of views from the Institute.

EDITORIAL COMMITTEE

CLIMATE CHANGE IN THE MEDIA

REPORTING RISK AND UNCERTAINTY

JAMES PAINTER

REUTERS
INSTITUTE for the
STUDY of
UNIVERSITY OF
OXFORD JOURNALISM

I.B. TAURIS
LONDON · NEW YORK

Published by I.B.Tauris & Co. Ltd in association with
the Reuters Institute for the Study of Journalism, University of Oxford

Published in 2013 by I.B.Tauris & Co. Ltd
6 Salem Road, London W2 4BU
175 Fifth Avenue, New York NY 10010
www.ibtauris.com

Distributed in the United States and Canada Exclusively by Palgrave Macmillan
175 Fifth Avenue, New York NY 10010

ISBN: 978 1 78076 588 4

A full CIP record for this book is available from the British Library
A full CIP record is available from the Library of Congress

Library of Congress Catalog Card Number: available

Typeset by 4word Ltd, Bristol
Printed and bound in Great Britain by TJ International Ltd, Padstow, Cornwall

Contents

Executive Summary

Politicians, scientists, and policy makers are increasingly using the concept and language of risk in a context of uncertainty to frame what is probably the greatest challenge this century, human-induced climate change. As much of the debate about climate change concerns the future, it inevitably involves degrees of uncertainty about the timing, pace, and severity of possible impacts, as well as the options for managing and avoiding them.

But uncertainty can be an obstacle to decision making. And scientific uncertainty is often misunderstood, particularly by the general public, and misinterpreted as ignorance. Many people fail to recognise the distinction between 'school science', which is a source of solid facts and reliable understanding, and 'research science' where uncertainty is engrained and is often the impetus for further investigation.

One of the arguments in favour of using the language of risk is that it shifts public debate away from the idea that decisions should be delayed until conclusive proof or absolute certainty is obtained (a criterion that may never be satisfied), towards timely action informed by an analysis of the comparative costs and risks of different choices and options (including doing nothing).

Another is that risk is an essential part of everyday experience, including the worlds of insurance, health, and investment. Many people have to deal with it daily and manage it in different ways: most people in the developed world take out house insurance against the low probability, very high impact event of a fire. Patients are increasingly familiar with the concept of the risks and benefits of different health treatments (though they rely on trusted intermediaries to help them to navigate the risk). And some of the risk assessments people make are on the same timescale as possible climate impacts – for example, taking out a pension policy into which they pay for 40 years.

There is also a growing body of literature suggesting that risk language may be a good, or at least a less bad, way of communicating climate change to the general public. Many argue that when compared to the messages of disaster or uncertainty that often surround climate change, risk is far from being a panacea, but it does offer a more sophisticated and apposite language to have the discussion in and a more helpful prism through which to analyse the challenge.

Risk can range in meaning from a broad sense of a possible adverse impact, to a narrow, more technical sense of assigning probabilities or confidence levels to different outcomes. Unlike previous analyses of media treatments of climate change, this study included an 'explicit risk' frame capturing a narrow sense of risk and compared it with three other narratives or messages often found around climate change: disaster (or implicit risk); uncertainty; and opportunity.[1]

An examination of around 350 articles in three newspapers in each of the six countries with a combined circulation of at least 15 million[2] showed that the dominant messages that readers receive were predominantly ones of disaster or uncertainty. The language of risk (and of opportunity) was much less prevalent. This was true for most of the climate change stories examined, and across the different media and political contexts of the six countries, and the range of newspapers.

Our other main findings were:

- The disaster/implicit risk frame was present in more than 80 per cent of the articles, making it the most common frame. For coverage of three reports by the Intergovernmental Panel on Climate Change (IPCC), it was present in over 90 per cent of them. It was also the most salient (in the headline or first few lines) with 44 per cent of the articles containing the frame, more than twice the next most common frame. It was also by some margin the most dominant tone of all four frames with well over half the articles containing it.[3]
- Uncertainty was the second most common frame after the disaster frame. It was present in nearly 80 per cent of the articles. However, it was less salient, and much less frequently a dominant tone.
- Opportunity was the third most common frame, being present in 27 per cent of the articles. However, these were overwhelmingly the opportunities from not doing anything about reducing greenhouse gas emissions. Only five articles (less than 2 per cent) in the total

sample contained a mention of the opportunities from switching to a low-carbon economy.

- Explicit risk was the least present of the four frames, and the least salient. It was the dominant tone in just three articles, although it was often combined with other frames. Its presence would have been more marked if the articles covering the three IPCC reports had included an explanation of the concepts of likelihood and confidence levels – only 15 per cent of the articles did so.

- Journalists follow the prompts from scientists and their reports: 70 per cent of the articles covering the IPCC reports, and nearly 60 per cent of all the articles in the sample, included quotes from scientists or scientific reports expressing some variant of the disaster/implicit risk frame. Nearly half of all the articles included a quote that indicated some manifestation of uncertainty.

- The one major exception to the lack of variation between the six countries was the presence of uncertainty through sceptical voices: Australia had the highest number of articles in the sample with sceptics in them and the highest percentage of articles, followed by the USA.

The implication of this study is that the language and concept of risk is not yet as embedded into climate change coverage as other strong narratives. This may well change, as the coverage of the 2012 IPCC report on extreme weather events shows. Although this report was covered in far fewer articles than the 2007 IPCC reports, the explicit risk frame was present in half of them, and was often a dominant tone.

The continuing appeal of the disaster frame is to be expected, and is in line with other studies, confirming that journalists are generally attracted to gloom and doom stories. But journalists are going to become more exposed to the language and the concept of risks in covering climate science. Numbers and probabilities are likely to become more important in the coming years: as climate models become more powerful and sophisticated, their potential to quantify uncertainties and generate probabilistic climate projections will be enhanced. In other words, uncertainty will be measured in more helpful ways as a basis for making decisions, even though it will not be eliminated. Giving ranges of probabilities and reliability or confidence levels will be an essential part of that.

Another reason to understand probabilities better is because a major area of research by the climate science community is focused on attributing, in terms of changes in risk, the role that human-induced global warming may have played in making individual extreme weather events more or less likely.

Bridging the gap between specialists and journalists is becoming more difficult at a time when specialist journalism is in decline in many Western countries, in part due to the problems facing the business model of print media. It is a worrying trend that most journalists are now generalists yet they have to cover highly specialised areas of risk in finance, health or the environment. It is an urgent task to encourage and help journalists to bridge the gap. Specifically, journalists will need to be able to handle risk, numbers, and probabilities better in order to help a more constructive narrative about climate change than doom and gloom or uncertainty.

Amongst the study's recommendations are:

- More familiarity and training for journalists about numbers and probabilities will improve coverage of climate risks.
- More scope for inclusion in website articles of details and discussion about how uncertainty can be quantified and given a confidence level.
- More (judicious) use of info-graphics to illustrate the concepts of risk and other aspects of climate change.
- More use of probabilistic forecasting in public weather forecasting on television.
- More resources for the IPCC to be able to communicate effectively around its reports and to respond to media enquiries quickly.
- Scientists should stress early on during media interviews where there is broad consensus about climate science, and then later on where there are degrees of uncertainty. They should also try to explain that uncertainty does not usually mean ignorance.
- Using the language of risk in the context of uncertainty can be a helpful way of presenting the problem to policy makers; but more research is needed about the effect on the general public of different types of risk language to test when it is effective, under what circumstances, with what groups, and with what metaphors.

Acknowledgements

The Reuters Institute for the Study of Journalism (RISJ) is very grateful to the following organisations for funding this study:

The European Climate Foundation

The Grantham Research Institute on Climate Change and the Environment at the London School of Economics and Political Science

Klif, the Climate and Pollution Agency in Norway, a directorate of the Environment Ministry

The RISJ would also like to thank Green Templeton College for funding a conference on reporting risk and uncertainty held in November 2012, which provided invaluable material for this study.

Acronyms and Abbreviations

ABC	Australian Broadcasting Corporation
ACIJ	Australian Centre for Independent Journalism
BBC	British Broadcasting Corporation
CCRA	Climate Change Risk Assessment
CIA	Central Intelligence Agency
CJR	*Columbia Journalism Review*
CRED	Center for Research on Environmental Decisions
EPA	Environment Protection Agency
EU	European Union
GHG	Greenhouse gas
GTC	Green Templeton College
GWPF	Global Warming Policy Foundation
IEA	International Energy Agency
IPCC	Intergovernmental Panel on Climate Change
LSE	London School of Economics
NGO	Non-Government Organisation
NICE	National Institute for Clinical Excellence
NOAA	National Oceanic and Atmospheric Administration
NPCC	New York Panel on Climate Change
NYT	*New York Times*
RISJ	Reuters Institute for the Study of Journalism
SREX	(IPCC Report on) Managing the Risks of Extreme Events and Disasters to Advance Climate Change Adaptation
TOI	*The Times of India*
UCL	University College London
UCS	Union of Concerned Scientists
UKIP	United Kingdom Independence Party

UMP	Union for a Popular Movement (Union pour un Mouvement Populaire)
UNEP	United Nations Environment Programme
UNFCCC	United Nations Framework Convention on Climate Change
WG	Working Group
WMO	World Meteorological Organisation
WSJ	*Wall Street Journal*

Tables and Figures

Tables

Figures

Preface

First and foremost we are very grateful to the researchers for the individual country studies, who were consistently patient, resilient, and insightful in the face of constant demands on them: Lyn McGaurr, Research Associate in the School of Social Sciences at the University of Tasmania, and Libby Lester, Professor of Journalism, Media, and Communications in the same School; Anu Jogesh, Research Associate with the Centre for Policy Research, New Delhi; Toussaint Nothias, PhD student and Teaching Assistant at the Institute of Communications Studies, University of Leeds in the UK; Christian Bjørnæs, Senior Communications Advisor at the Center for International Climate and Environmental Research Oslo (CICERO), and Anja Naper, Research Assistant at CICERO; and Cassie Tickell-Painter, a researcher at Meteos, a think tank on health and the environment based in Oxford, UK.

A number of experienced environment journalists from the six countries we examined kindly found time to respond thoughtfully to questions from the researchers into their reporting of risk and uncertainty around climate change. They were, in Australia, Sarah Clarke of ABC and Ben Cubby of the *Sydney Morning Herald*; in France, *Le Monde*'s Stéphane Foucart and *Le Figaro*'s Marielle Court; in India, Meena Menon, the Mumbai Chief of Bureau for *The Hindu*; in Norway, Guro Tarjem, a reporter for NRK Radio; in the UK, Fiona Harvey, formerly of the *Financial Times* and now of the *Guardian*; and Alister Doyle of Thomson Reuters; and in the USA, Andy Revkin, the former *New York Times* environment correspondent and now the author of the Dot Earth blog.

The two official reviewers of the text, Rasmus Kleis Nielsen and Adam Corner, provided very helpful suggestions for changes, as did David Levy, Director of the RISJ. Detailed comments were kindly offered by Richard Black and Bob Ward. Vaclav Stetka helped with the design of the content analysis. The long list of others who helped in differing amounts was Teresa Ashe, Curtis Brainard, Ros Donald, Anna Gerrard-Hughes, Catherine Happer, Ian Hargreaves, Chris Hope, Christian Hunt, Patrick McSharry, Tim Palmer, Chris Rapley, Yves Sciama, Juan Señor, William Solecki, David Stainforth, Liz Stephens, Nick Stern, Sophia Tickell, and Maya Tickell-Painter. Sara Kalim and Alex Reid from the RISJ have offered their customary, but greatly valued, support.

As always, all errors of judgement and fact are mine.

1

Introduction –
'Even Ostriches Need Third-Party Insurance'

2012 was an extraordinary year for extraordinary weather.[1] The extreme became commonplace. It did not seem to matter much where you lived on the planet. The UK experienced some of the most unusual weather on record as the driest spring for a generation gave way to the wettest recorded April to June in a dramatic turnaround never documented before. By the end of December, England had suffered the wettest year on record, the whole of the UK the second wettest.

Many other parts of Europe had their worst cold snap in decades; in China the average winter temperature hit its lowest in 28 years; warmer temperatures in the Arctic were one factor behind a record low level of sea ice cover in September; Australia suffered its hottest summer since records began in 1910; the north-east of Brazil had its worst drought in decades; and in September, Nigeria experienced its worst flooding in 50 years.

In the USA, 2012 was particularly extraordinary for a series of weather extremes, many of them unparalleled in recent American history. The average temperature for the year was a full degree warmer than the previous record set in 1998. It was, in the words of a senior climate observer, 'a huge exclamation mark at the end of a couple of decades of warming' (Borenstein, 2013).

The area of contiguous USA (the 48 states not including Alaska or Hawaii) suffering drought conditions peaked at around 62 per cent in July – the largest area to be affected since the infamous Dust Bowl drought of December 1939. That July was the warmest on record in the USA for any month since records began in 1895.

Hurricane Sandy made landfall on 29 October near Atlantic City in New Jersey. At that moment, it registered sustained winds of 80 mph, and a central minimum pressure of 946 millibars, the lowest pressure ever recorded along the north-eastern coast. Parts of New York City harbour registered a high water level of nearly 14 feet, beating the previous record by three feet. Sandy's storm-force winds covered more than 940 miles of the north-eastern US seaboard. It was the largest Atlantic hurricane on record, as measured by diameter.

Sandy's political impact may not have been as significant as the tens of billions of dollars of damages, but it was hardly negligible. President Obama could be seen being presidential and sympathising with victims, while the Republican challenger Mitt Romney struggled to find a role. But of central importance to this study was the reaction of the New York mayor, Michael Bloomberg, who threw his weight behind Obama because of Romney's failure to back climate change measures. The *Guardian* put these words from Bloomberg on its front page on 2 November:

> *Our climate is changing. And while the increase in extreme weather we have experienced in New York and around the world may or may not be the result of it, the risk that it might be – given this week's devastation – should compel all elected leaders to take immediate action. (MacAskill and Goldenberg, 2012)*

The three concepts of climate change, uncertainty, and risk were all woven together in this short quote. The dominant sentiment was remarkably consistent with what most climate scientists would maintain about climate change and its link to the individual weather extremes described above. Despite recent advances in attribution modelling (see for example Otto et al, 2012; Pall et al, 2011; Stott et al, 2004), most of them regard it as misguided to attribute single weather events – and particularly tropical storms – too closely to human-made climate change. But it was Bloomberg's representation, or, to use the terminology of social science, his 'framing' of the climate change problem that was particularly interesting for those large groups of academics, policy makers, psychologists, and environment groups who have long grappled with the fiendishly difficult challenge of communicating climate change.[2] Bloomberg was linking climate change to current extreme weather events, rather than future ones, or in other words he made the threat immediate rather than future.

There is a growing sentiment amongst several experts that placing more emphasis on the climate change challenge as risk may be a helpful tool in framing or communicating the uncertainties around it. These advocates of a 'risk-based' approach argue that – in certain circumstances – such a framing can give policy makers more clarity about options and the process of making decisions about them. Some also argue it is less of an obstacle to public understanding, engagement, and behaviour change than other ways of portraying climate change, although this view is more contested.

In 2010, Bloomberg's office wrote a report outlining the multiple threats posed to New York by climate change, called 'Climate Change Adaptation in New York: Building a Risk Management Response'.[3] Bill Solecki is the co-chair of the New York Panel on Climate Change (NPCC) which wrote the report and an advisor to Bloomberg. He explains that 'when Bloomberg endorsed Obama like that, he was following the same logic we have been following at the NPCC. Our basic approach has been that it is hard to say that any one extreme event is climate change, but it is clear that the environmental baseline of the city is changing. So it is prudent to make the city more resilient for all types of climate risk – both present and future.'[4]

Bill Solecki is one of several experts who now portray the climate change problem as one of risk in a context of uncertainty – the risk that the continuing rise in greenhouse gas (GHG) emissions from human activities may well exacerbate weather extremes and cause widespread and harmful impacts. As has been well-documented, and as we shall review in Chapter 2, there are large areas of uncertainty that surround the climate, certainly about the past but more pertinently about the future. The planet's climate system is immensely complex and difficult to understand fully; the computer models used by natural scientists give multiple ranges of future temperature increases and potential outcomes; and social scientists are cautious about the possible pace, timing, and scope of the social and economic impacts from higher concentrations of GHGs in the atmosphere.

However, some eminent climate scientists and economists have long argued that although there are these important uncertainties surrounding the future of our climate, a better way of looking at the problem is to emphasise the risks. The US climate professor Stephen Schneider of Stanford University, who spent much of his life engaging in public discussion about climate change and thinking about the role of the media, was one of the first to employ the everyday concept of risk.

He would ask his audience, 'How many of you own a home?' Many of his (often well-off) audiences would put up their hands. 'How many of you have had a house fire?' Very few would put up their hands (typically it is much less than 1 per cent of households, although 1–2 per cent in California). He would then ask, 'How many of you have fire insurance?' Most people would put up their hands.

He would then proceed to point out that many people are happy to manage risk at a personal or household level, even when they are acting on a very low chance (less than 1 per cent) of a negative impact. But with climate change, he argued that the risk to the planetary life support system was much higher, and yet some sceptics were saying there was not enough certainty to take action.[5]

In the UK, the Conservative former secretary of state for the environment, John Gummer, uses the same analogy with the house insurance market. He called a talk in February 2013 at Oxford University 'even ostriches have third-party insurance',[6] and drew parallels between how people deal with questions of insurance and how governments should deal with climate risks. He said that most British people are not ostriches and take out house insurance, but sceptics 'were constantly saying that because there was no absolute certainty, we should do nothing'.

And in Australia, the leading climate scientist Professor Will Steffen uses a different form of risk language in describing the link between global warming and Australia's recent records in weather extremes. He told the *Sydney Morning Herald* in March 2013 that 'statistically, there is a 1-in-500 chance that we are talking about natural variation causing all these new records. Not too many people would want to put their life savings on a 500-to-1 horse' (Siegel, 2013).

The seminal 2006 Stern report on the economics of climate change was probably the first to explicitly represent the climate policy problem as one of decision making in a context of uncertainty and risk. Its author, Lord (Nicholas) Stern, still argues that climate change is all about risk management, albeit on a colossal scale:

> *The main point is that this is all about risk management in some shape or form. Do we want to play Russian roulette with one bullet in the barrel or two? And even if we can't be that precise about one or two we can say that we can, through sensible action, dramatically cut the risks. And delay is dangerous because of the ratchet effect of a flow-stock process and the lock-in of capital and infrastructure. Thus those who favour inaction have to say that they know the risks are very small – agnosticism does not make their case.[7]*

In his report, Stern famously recommended that it was necessary to invest 1 per cent of the world's GDP annually for the next few decades to move from a high-carbon to a low-carbon economy (although he now says it is more like 2–3 per cent because of the greater costs involved in keeping GHG emissions to below 450 ppm). Those who baulked at the tens of billions of dollars this would entail were reminded that the global insurance industry, excluding life insurance, is worth 3.5 per cent of global GDP. As one Australian commentator observed at the time,

> If the world is prepared to pay the equivalent of 3.5 per cent of its total annual output to guard against the possibility of all sorts of risks that, in any one year for any one client, are quite remote, such as fire and theft, then the prospect of paying a 1 per cent premium to protect against a catastrophic global event seems entirely reasonable. (Hartcher 2007)

A number of more recent reports have explicitly introduced the concept of risk into their titles and focus. Two stand-out examples of this are the March 2012 report by the world's most important climate science body, the IPCC (the Intergovernmental Panel on Climate Change), on extreme weather events (Managing the Risks of Extreme Events and Disasters to Advance Climate Change Adaptation or SREX report), and the UK Climate Change Risk Assessment (CCRA) of January 2012, which was the first of its kind in the UK.[8] For example, the CCRA pointed out that there were a large number of uncertainties surrounding climate models leading to a wide range of possible results, but stressed (unusually) that not only were there some opportunities from climate change (new shipping lanes through the Arctic and fewer cold-related deaths in winter) but also multiple risks from flooding, heatwaves, and water shortages. Indeed, the CCRA press release mentioned the word 'risk' 19 times.

The portrayal of the climate change challenge as managing risk is certainly a growing trend, particularly in policy making circles, and this is just one reason why it is important to study the way the media report risk and uncertainty. There are several others.

Greater risk, but (in some areas) greater uncertainty too

In late 2012 a swathe of reports from the World Bank, the CIA, the International Energy Agency (IEA) and the United Nations Environment Programme (UNEP) all highlighted the growing possibility of a 3-degree or 4-degree warmer world by the end of the century, and the greater risks that this would entail (Clark, 2012). For example, the IEA warned in November 2012 that the world is likely to build so many fossil-fuelled power stations, factories, and inefficient buildings in the next five years that it will become impossible to hold global warming to safe levels, which, in the judgement of governments, means less than 2°C.

In the same month, the World Bank's president, Jim Yong Kim, made an urgent plea for action to address the 'devastating' risks of climate change while launching the Bank's report 'Turn down the heat'.[9] The report detailed the impact of a world hotter by 4°C by the end of the century, which the Bank described as 'a likely scenario under current policies'. It said that 'extreme heat waves would devastate broad swathes of the earth's land, from the Middle East to the United States. The warmest July in the Mediterranean could be 9°C hotter than it is today – akin to temperatures seen in the Libyan Desert'.

Proponents of the tipping point approach argue that one of these points may have already passed with the (then) record Arctic sea ice melt of 2007 (see for example Marshall, 2013). The tipping point approach holds that a system can exist in several more or less stable states, and that when a certain threshold is reached, it 'tips over' from the state we see now into another one. Furthermore, as there is an interconnected web of tipping points, when it comes to impacts or consequences of climate change, a small change can unleash a big change which may be unstoppable. For example, when the Arctic sea ice flips into a new, less stable state, this may push the planet quickly into another tipping point – the thaw of a vast expanse of the Siberian permafrost.

However, at the same time that the risks were being laid out with greater urgency, in an apparent paradox some prominent experts on climate change also stressed that, as researchers dig deeper into the field of climate science, more uncertainties about some aspects are being, and will be, uncovered. For example, one widely-quoted article in *Nature* written by a leading climate scientist argued that despite the advances in our knowledge about climate science, the 2013–14 report by the IPCC would have a *greater* amount of uncertainty in some of its predictions and projections, which could 'present a major problem for public understanding' (Trenberth, 2010). Indeed, a leaked draft of the first section of the report suggested that the IPCC was changing some of its projections about droughts, tropical cyclones, and ocean circulation (Marshall and Pearce, 2012). And the slowdown in the increase in global mean surface air temperatures since 1998 has led to heated debate in the media about climate sensitivity, or the amount of global surface warming that will occur if the concentration of CO_2 in the atmosphere is doubled (see for example *Economist*, 2013; Rose, 2013).

Public understanding of scientific uncertainty

As we shall see in Chapter 2, the general public are often unaware that many areas of science involve uncertainty whereas, for a scientist, uncertainty is often seen as something positive which can prompt further research. As the former director of the British Antarctic Survey, Professor Chris Rapley, observes, 'there is a fundamental disconnect between scientists and non-scientists, as the general public associate science with things we know about with a good deal of certainty like gravity, DNA or the Higgs-Boson. But a lot of science is about uncertainties'.[10] Even the word 'uncertainty' is frequently interpreted by the lay public to imply complete, rather than relative, lack of knowledge – or in other words, ignorance.

In the realm of climate science, the gap between public understanding and what most mainstream climate scientists believe has long been documented. Various opinion polls suggest that the public in many countries are confused about where there is broad consensus amongst scientists on aspects of climate science (for example, the warming trend and human influence), and where there is much less consensus or more uncertainty (the timing and extent of impacts, and what policy actions to adopt). For example, in the USA, one widely-respected survey in 2012 found that 54 per cent of Americans believe global warming is caused mostly by human activities, which compares with the 95 per cent plus of climate scientists who think it is.[11] The same survey suggested that only 44 per cent of Americans believe most scientists agree that global warming is happening. It is a surprising figure, and one that is replicated in other countries.

One of the many reasons for this mismatch is the way climate sceptics of different persuasions jump on the uncertainties, in order to cast doubt on parts, or even the whole edifice, of climate science. There is considerable evidence that in recent years climate scepticism has been on the increase in the Anglophone media, and in public opinion in some countries (Painter, 2011). On the other hand, green NGOs can downplay uncertainties as an obstacle to public engagement, for example by attributing an extreme weather event too specifically to man-made global warming. So, the misrepresentation of scientific uncertainty can have an adverse effect on public understanding.

The communication of uncertainty and risk

In recent years, there has been a proliferation of academic studies, many written by psychologists or communication specialists based on work with focus groups, looking at the different ways uncertainty and risk are communicated, and the effect of such messages on public understanding, engagement, and behaviour change (for an overview, see Moser, 2010: 37). In the specific area of climate science, the way the public receive, and react to, the information is in part a product of the way it is framed, either in the media or other forms of communication. For example, as will be discussed in Chapter 2, stressing uncertainties can lead to a diminishing of the problem in the minds of the reader or viewer. If scientists constantly talk about uncertainty, often the response of the listener is not necessarily apathy but lapsing into an unhappy situation of not knowing how to proceed, and therefore discounting or dodging the problem. Or some can even get angry when scientists don't know things. Some academics argue that climate change needs to be re-framed away from the technical uncertainties in the science and more towards the risks to society, and this change of emphasis can enhance engagement and understanding (Pidgeon and Fischhoff, 2011).

So a lot of research has already been carried out on the uncertainties and risks around climate change, the public understanding of these issues, and the way they are communicated and can be received. What's often missing is an understanding of what newspapers and journalists do when they cover risk and uncertainty, why they do it that way, whether they have the training and tools to understand scientific uncertainties, and whether they are sufficiently numerate to understand different concepts of risk.

The role the media play is crucial. After all, most people get most of their information about science from the media – maybe as much as 80 per cent of the population in the UK.[12] Of course, scientific news stories are often inherently specialist, containing ideas and language that are unfamiliar to most of the lay public. Risk and uncertainty are particularly difficult concepts to convey to the public and can make the journalists' job difficult. Moreover, the journalists are often being targeted by lobby groups of all persuasions to amplify or underplay uncertainties around climate science, which has at times contributed to a lack of nuance or complexity in reporting.

But the task of getting it right is an urgent one, going beyond climate science. Two cases taken from 2012 of the media's reporting of scientists' statements involving risk and uncertainty, the first around cancer risks, the second around the risk of a major earthquake, illustrate both the challenges and the consequences of not getting it right. Although the cases are very different, they both highlight the essential question of what constitutes responsible reporting.

In the UK, studies have shown how the risk from eating red and processed meat in general, and bacon sandwiches in particular, has been exaggerated in the media, in part because of the way the press releases are written (Riesch and Spiegelhalter, 2011). For example, the *Daily Express* in a front page report on 13 January 2012 reported the potentially scary piece of information that a 'daily fry-up boosts cancer risk by 20 per cent'.[13] The report was based on a study published in the *British Journal of Cancer*. At no point did the *Express* article point out that the 20 per cent increase in the chance of getting pancreatic cancer was from a very low base, and that the number at risk went up from 5 in 400 to 6 in 400 as a result of eating one sausage or two pieces of bacon a day. Six is a 20 per cent increase from five, so the relative risk did increase, but the absolute risk remained small. It was a classic example of not reporting the denominators, as we shall discuss further in Chapter 4.

A different set of issues arose from the case in Italy of three seismologists, two engineers, a volcanologist, and a public official being sentenced to six years in jail in October 2012 in L'Aquila for falsely reassuring or 'over' reassuring the general public about the likelihood of a major earthquake occurring in March/April 2009. For the media, the issue was in part whether they accurately reported the caveats and qualifications that the scientists said they included. For the scientists, the main issue was, as highlighted by the UK academic Brigitte Nerlich (2012), how scientists should:

> navigate between the Scylla of being open about uncertainty and the Charybdis of public and political expectations regarding pronouncements of certainty, as well as, and perhaps more importantly, between the rock of scare mongering and the hard place of 'complacency mongering'.

This study aims to travel a small way towards filling some of the gaps about the media's reporting of future risks. Before taking a detailed look at this in Chapters 5 and 6, we review in Chapter 2 how researchers from different disciplines define uncertainty, and how this can differ from what the public understand by it. We then look more closely at the uncertainties around climate science, the way the IPCC communicates them in terms of likelihoods and degrees of confidence, and the exploitation of uncertainties by climate sceptics.

In Chapter 3, we show how risk is part of our everyday experience, and particularly in the insurance, investment, and health sectors. We then outline the arguments in favour of using risk language and metaphors for the climate challenge, and assess its effectiveness in helping policy makers to make better-informed decisions, or in helping understanding and engagement from the general public. We end this chapter with an account of what experienced journalists think about the advantages and disadvantages of using risk language.

In Chapter 4, we review what other studies have shown about the way the media report uncertainties and risks in general, and also about climate science. We also discuss here what might help journalists and the general public understand risk better, such as more probabilistic weather forecasting on TV (in some countries) and the judicious use of info-graphics.

With all of the above as context, we then focus in Chapter 5 on the ways uncertainty and risk are reported in the (mostly elite) print media in six countries (Australia, France, India, Norway, the UK, and the USA). We carry out content analysis of the ways three newspapers in each country have covered two aspects of the possible future of the planet as a result of global warming: projections of global temperatures and of Arctic sea ice melt. The advantage of these two topics is that they both involve uncertainties, risks, and opportunities. Specifically, we examine the reporting of the IPCC's first two reports in 2007, the IPCC's SREX report of March 2012, and the reporting of Arctic sea ice decline.

We tested the relative presence in around 350 articles of four main 'frames', namely uncertainty, disaster/implicit risk, explicit risk, and opportunity. A full description of these frames can be found in Chapter 5 but, given the focus of this study, we looked closely at the 'explicit risk' frame which includes numerical probabilities, the use of the word 'risk', or the inclusion of everyday concepts or language like insurance, betting or not taking unnecessary risks (the precautionary principle). We tested the relative presence of these four frames, and give the aggregate results for all six countries, including any significant country differences. In Chapter 6, we give the social, political, and media context for each country followed by a detailed analysis of the results. Finally in Chapter 7, we draw out some conclusions and summarise some of the challenges and recommendations for scientists and journalists that arise from this study.

2

When Uncertainty Is Certain

The words 'uncertainty' and 'risk' are used in different ways in different academic disciplines, which in turn is often quite different to the way the public understand them. There are various definitions and various typologies for each of them (Spiegelhalter and Riesch, 2011), but most experts usually start by making a distinction between uncertainty and risk. Risk is often differentiated from the various forms of uncertainties in that the likelihood of something (usually adverse) happening can be quantified by probabilities. Sometimes risk is used to refer just to the likelihood of harm, and at other times it is used to refer to the likelihood of harm multiplied by the severity of the consequences. This is called risk in a narrow or statistical sense to distinguish it from the colloquial sense of the possibility of an adverse impact. We shall return to a discussion of this in the next chapter.

The general public are generally not familiar with the different definitions and types of uncertainty recognised by experts. For many scientists and engineers, 'uncertainty' means 'that in a particular situation more than one outcome is consistent with our expectations, and it is often expressed by giving a margin of error with every measurement' (Pielke, 2007: 55). As we shall see, the IPCC offers detailed guidelines for climate scientists on the consistent treatment of uncertainty across all its reports, which involve two key concepts of the degree of confidence in a finding, and quantified measures of uncertainty in a finding expressed as a probability (Mastrandrea et al, 2012).

Uncertainty can result from lack of information or from disagreement over what is known or even knowable. While many sources of uncertainty exist, it is helpful, even for the lay person, to distinguish between uncertainties that are not known in practice but could in principle be known or reduced ('epistemic uncertainties'), and uncertainties that can never be known or will not go away ('aleatory uncertainties'). This latter type of uncertainty is associated with unavoidable unpredictability or chance, as for example in the throwing of dice. This sort of distinction can be very helpful for those modelling future risks. A modeller can be clearer as to which uncertainties have the potential of being reduced by gathering more data or by refining models (epistemic ones), compared to those which cannot be reduced (aleatory ones). But, even then, some modellers add a sub-category of epistemic uncertainties: uncertainties that will go away and it can be predicted how quickly they will go away.[1]

The American professor of geophysics Henry Pollack was one of the first popular science writers to explore the differences between how scientists and lay people view the uncertainties in science, and the role the media play in communicating them. In his 2003 book *Uncertain Science... Uncertain World*, he lucidly explains how non-scientists often equate science with certainty and not uncertainty, and can become surprised, impatient or discontented when scientists cannot show a high level of certainty in their understanding of complex natural systems (Pollack, 2003: 6). He cites the two examples of the public reaction in 2001 to the discovery of anthrax spores in public buildings in the USA and the spread of foot and mouth disease in the UK.

David Stainforth from the London School of Economics (LSE) uses the two concepts of 'school science' and 'research science' to explain why this might be so. His view is that

> *perhaps people's perception of science originates from what they were taught at school. This 'school science' is the source of solid facts and reliable understanding. [It] is very different to 'research science'. The former is about communicating what we already understand, the latter about developing and expanding our understanding. In research science new results and interpretations are continually developed. Disagreements and debate are common.*[2]

As adults too, people see and experience certainties at work in science. On a practical level, science diagnoses illnesses and has developed and tested cures, which usually work. Men did land on the moon; rivers do flow downhill, and so on.

So for a scientist, uncertainty is often seen not as something negative but as a positive stimulus to further research (or future funding). For a lay person it can mean something very different. In some contexts, just the word 'uncertainty' is frequently interpreted by the general public to imply complete, rather than relative, lack of knowledge (Shuckburgh, Robison and Pidgeon, 2012; Somerville and Hassol, 2011). Or when scientists acknowledge they do not know everything about for example the spread of a communicable disease, the public translates this to mean they do not know *anything* about it. According to Henry Pollack, this 'leads to a loss of public credibility […]. A by-product […] is an all-too-frequent willingness of the general public to entertain flimsy pronouncements from kooks, charlatans and marginal skeptics' (Pollack, 2003: 7).

The effective communication of uncertainties has been the subject of a large body of work, partly because it has become so relevant to the communication of health risks, particularly from possibly contaminated food, to the general public. The American risk expert Peter Sandman writes that it is a real dilemma for officials as to what they should say when you have uncertain risk information to impart. His starting point is that uncertainty should be explained and proclaimed as much as possible, and not hidden (Sandman and Lanard, 2011). He writes that 'People hate it when you express the uncertainty. They hate it so much that they tend to miss it. Unless you aggressively, unambiguously proclaim your uncertainty, people are likely to imagine that you're a lot more certain than you are.'

Whereas the risks from eating contaminated food are immediate, many of the worst impacts of climate change are probably a long way off in time and space. In the world of climate science, studies done by psychologists or communication scientists strongly suggest that stressing uncertainties (either directly quoted from scientists or as general statements) can lead to a diminishing of the problem in the minds of the audience, and act as an obstacle to action. Specific studies in the USA suggest that an early message about uncertainty 'gives listeners the permission to dismiss or turn attention away from what follows' (Moser and Dilling, 2004). Some of the focus group participants in a 2012 UK study about climate change were 'looking for definitive statements and were very sensitive to the use of words such as "could", "may" and "suggest"' (Shuckburgh, Robison and Pidgeon, 2012: 30). Some were left frustrated or even angry with uncertainty. Or participants responded by saying scientists were 'sitting on the fence', or that a news report was 'lacking facts' or not being 'conclusive'. One respondent said that reading such articles felt 'like a waste of time that you will never get back if you're reading hypotheses all day long'.

The study also found that the media's use of such phrases as 'scientists are uncertain that …' had a different impact to ones like 'scientists fear that, or say there is a risk that …' As Professor Chris Rapley sums it up, 'detailed descriptions of uncertainty can drive listeners to feel anxious and fearful. A natural response to these feelings is to apply coping strategies, such as disavowal and denial to get rid of the unpleasant feelings.'[3]

Uncertainty in climate science

The wide range of uncertainties surrounding climate science have prompted many experts to take up the challenge of trying to map them, to compare them with the more certain aspects, and to use a language that can make them understandable to their (often different) target audiences. Most areas of climate science fit much more into the category of 'research science' than 'school science' and, just like other areas of complex research science, uncertainty is a feature of many aspects of climate science. It may be able to be reduced in some areas, but it is unlikely to go away completely. But for the general public educated in 'school science', this uncertainty may be something they do not readily associate with science.

So what do scientists tell us about the degrees of certainty about different aspects of climate science? In broad brush terms, there is more certainty about the past, but (unsurprisingly) much less certainty about the future. So, there is almost universal agreement that the average surface temperature of the Earth has warmed in recent decades, and that the warming is mainly a result of human-driven emissions of GHGs. There is much less certainty about how quickly the planet will warm up, and about the timing, scope, and location of the impacts of this warming. However, some scientists would say that even this characterisation of the uncertainties is problematic. They would argue that there is also high certainty that if we get more than 2 or 3 degrees of warming, the impacts will be very disruptive of the global climate and severe (possibly very severe) for many people living in different parts of the world – even if we cannot be certain about how exactly these impacts will unfold.[4]

There are many ways to present the uncertainties around climate science, but Table 2.1 is an attempt to show in simplified form where the main areas of high certainty and significant uncertainty are found. The table aims to capture three areas of high certainty and four areas of significant uncertainty. The latter is divided into (i) How much warming

Table 2.1   Certainty and Uncertainty in Climate Science

Level of certainty/uncertainty		Sources or types of uncertainty
High certainty	The average temperature of the earth has warmed about 0.8°C since the beginning of the twentieth century	
High certainty	The warming is largely a result of human activity (the emission of GHGs)	
High certainty	If the warming is more than 3°C, the warming will be very disruptive of the global climate, and will have predominantly negative impacts on humans, plants and animals	
Significant uncertainty	1) How much warming will take place, over what period?	Strength of feedback, especially linked to behaviour of clouds; carbon cycle; dampening effect of human aerosol emissions
Significant uncertainty	2) How robust are climate models?	Understanding complex processes governing climate dynamics; carbon cycle; dampening effect of human aerosol emissions; trends in precipitation
Significant uncertainty	3) How will the Earth respond to warming temperatures?	Timing and magnitude of ice sheet melt affecting sea level rise; tipping points
Significant uncertainty	4) What will governments and society do in response?	What policy actions will be taken, when, affecting future emissions?

Sources: Harrison (2013) and websites referenced in notes 6 to 9

will take place, over what period? (ii) How robust are climate models? (iii) How will the Earth respond to warming temperatures? and (iv) What will governments and society do in response?[5] The final column includes some of the main sources or types of uncertainties in each category.

The table could have been expanded to include many other areas, sources or types of uncertainty. As we shall see in Chapter 6, there is a range of uncertainties about Arctic sea ice melt, and there are also debates about whether hurricanes will increase in intensity and frequency, the precise links between some extreme weather events and global warming, and the projection of ranges, timing, and location of sea level rises. And within some of the topics that have been included, there are disputes as to whether the computer models are (or will be in the near future) reliable enough for informed regional or local level decision making (Schiermeier, 2010).

Nor does the table distinguish between the 'epistemic' and 'aleatory' uncertainties. Indeed, some experts have argued that in order to help decision makers, scientists need to distinguish between the quantifiable and unquantifiable, namely the 'uncertainty that can be quantified by a range, either with statistics or based on "what if" assumptions, and uncertainty that cannot be quantified, due to methodological unreliability, recognised ignorance, or value diversity in the practice of climate science' (Petersen, 2012). Nor does the table differentiate between 'model-based' uncertainty (for example, a 20 per cent model confidence in an effect) and 'conflict-based' uncertainty (for example, two out of ten experts expressing confidence in an effect).

Faced with such a daunting range of representations of uncertainty, types of uncertainties, and sources of uncertainty, we chose to simplify. For the vertical axis of Table 2.1, we opted to use just two variables of certainty: a 'high degree of certainty' and 'significant uncertainty'. We could have used several other types of gradation and language. As Table 2.2 shows, a wide variety of different languages of certainty and uncertainty has been used by leading organisations, popular websites or well-known science magazines writing about climate change. What is significant is that, while the seven different types of publication included in Table 2.2 exhibit a high degree of consensus on what aspects of climate science are (nearly) certain or uncertain, the range of language they use to describe these is quite different.

For example, in the 2013 publication *Engaging with Climate Change* Professor Stephan Harrison uses just two variables, 'high certainty' and 'considerable uncertainty' (Harrison, 2013), whereas the website Skeptical Science run by the Australian scientist John Cook uses different phrases ranging from 'overwhelming evidence for' to 'significant uncertainties'.[6] The magazine *New Scientist* on its website divides the issue into 'what we

Table 2.2 The Language of Certainty and Uncertainty

Organisation/ publication	Language used		
	Certainty	High certainty–low uncertainty	Uncertainty
Engaging with Climate Change	'High certainty'		'Considerable uncertainty'
Skepticalscience.com	'Know', 'overwhelming evidence for'	'Almost certainly'	'Significant uncertainties'
New Scientist	'What we know'		'What we don't know'
Carbon Brief	'Strong evidence for'		'Less sure about'
Royal Society	'Wide Agreement'	'Wide consensus, but continuing debate and discussion'	'Not well understood'
Green Alliance	'Unequivocal evidence'/ 'high level of certainty'	'Very likely'	'Likely'
IPCC	'Virtually certain (>99% probability)'	'Extremely likely (>95% probability)' 'Very likely (>90% probability)'	'Likely (>66% probability)' 'Very unlikely (<10% probability)'

know' and 'what we don't know',[7] whereas the specialist website Carbon Brief also goes for the same simple dichotomy as a title, but in the body of the text uses the phrases 'strong evidence for' and 'less sure about'.[8] The most prestigious science organisation in the UK, The Royal Society, opts for three divisions in its 2010 publication 'Climate change: A summary of the science': 'Wide agreement', 'Wide consensus, but continuing debate and discussion', and 'Not well understood'.[9] And finally, the Green Alliance chooses a range from 'unequivocal evidence for' to 'very likely' and 'likely' in its briefing aimed mainly at UK politicians.[10]

Perhaps it is to be expected that the language of certainty and uncertainty is subject to such a wide range of treatments, given the different nature of the publications, target audiences, and aims of the organisation. So, the *New Scientist* is perhaps more likely to give priority to immediacy and understanding, whereas the Royal Society is more likely to use the language of scientists through such terms as 'consensus' and 'continuing debate'.

But it is possible to see some of the problems that arise from each of the different approaches. If the certainties and uncertainties are divided into 'don't know – do know', then this runs the risk of forcing every aspect of climate science into too Manichean an approach when much of it falls between the two extremes. 'Considerable' or 'significant' uncertainty prompts the question of how much uncertainty? Even if the phrase 'very likely' or 'likely' is used, most people would still want to know 'how likely'? The use of the language commonly expressed by scientists stressing 'consensus' or 'debate' certainly reflects the way the science community describes the certainties and uncertainties – they weigh up the evidence and tell you which of several possible answers has the most support. But that probably won't work for a mass audience.

The IPCC's approach

It was the subjective nature of people's interpretation of the word 'likely' that was a major driver of why the IPCC decided it had to find a better way of communicating the uncertainties around climate science. In his very readable account of how the IPCC came to use the concepts of likelihood and confidence levels in the 1990s, the US climate scientist Stephen Schneider explains that well-known decision analysts were called in to help the research community present their findings in a way that decision makers in government could find useful (Schneider, 2009). As one leading advisor explained to Schneider, even scientific experts differed widely on what they thought the probability of 'likely' meant in the context of cancer risk. The experts often disagreed by nearly a factor of ten.

For several months, Schneider and one of his colleagues acted as what he describes as 'uncertainty police' trying to get consistency across the different working groups and reports. They essentially followed a risk management approach, where possible outcomes were weighed numerically and given 'subjective probabilities' and, at the same time, levels of confidence were given for each possible outcome. This approach met with opposition from some physical scientists who did not like subjectivity and risk management, in part because for them a risk management approach felt too much like providing 'policy-prescriptive' information, rather than the 'policy-relevant' information they are asked to give. Some social scientists also felt that wider society, and not scientists, should choose how to take the risks after all the possible conclusions were reported, and not just the consensus ones. But it apparently proved popular with the governments who had to sign off on the IPCC reports and who had to make decisions based on the science.

As a result, the first two of the three IPCC reports of 2001 (known as the Third Assessment Report or TAR) included in their Summary for Policy Makers a detailed description of what such phrases as 'very likely' (90–99 per cent chance) and 'likely' (66–90 per cent chance) meant. In similar fashion, it was explained that 'very high' confidence in a finding meant 95 per cent or greater, 'high confidence' meant 67–95 per cent, and so on. A similar approach of combining numerical probability ranges and confidence levels was used in the IPCC reports of 2007 (known as the Fourth Assessment Report or AR4), although some of the ranges and language were changed. (See Appendix 2 for a fuller description.) The last row of Table 2.2 above shows how the IPCC approach compares to the language of other organisations.

The IPCC's combination of quantified confidence and probability levels has been widely praised by risk experts.[11] Many journalists have also expressed approval. (See box in Chapter 3.) The co-chair of the IPCC's WG-1, Professor Thomas Stocker of the University of Bern, is a strong advocate of the approach, but admits there are some minor problems with it, particularly around the term 'very likely' being too weak an assertion.[12] In a further refinement of the approach, the lead authors of the 2013–14 fifth assessment report were given key guidelines for dealing with uncertainty, in part to achieve a common approach across the various reports and disciplines which had been absent from previous reports (Mastrandrea, 2012).

However, as some researchers have pointed out, a more robust characterisation of the uncertainties, or yet more refinement of the taxonomy of uncertainties, is not enough to guarantee effective communication of them (Ekwurzel, Frumhoff and McCarthy, 2012). It should be stressed that the IPCC guidelines are aimed mainly at scientists, and IPCC reports are aimed primarily at decision makers, and not the media, the general public or academics. But there is some evidence for thinking that its use of some terms does not aid wider public understanding. The most widely-quoted criticism is based on research carried out by a group of psychologists working at the University of Illinois (Budescu, Broomell and Por, 2009). They concluded that a big discrepancy is often found between the meaning the IPCC had intended and the conclusion people actually drew. In particular, they found that the public consistently misinterpret the IPCC's probabilistic statements, and usually in a 'regressive' way, or, in other words, less extreme than intended by the authors of the reports. So for example, when the IPCC writes that 'many changes in the global climate system during the twenty-first century [...] would *very likely* be larger than those observed during the twentieth century', the IPCC intends this to mean more than 90 per cent probability, whereas the typical respondent interpreted this to mean a 65–75 per cent probability. The researchers also suggest that the IPCC method tends to convey levels of imprecision that are too high, which may in part explain why the public in many countries tend to underestimate the level of scientific agreement.

The authors of these studies and other academics suggest that whereas the public struggle with the ambiguity of verbal terms like 'unlikely' or 'probable' when they are used on their own, they find it easier to recognise uncertainties when the verbal terms are accompanied by the actual numbers giving the probability ranges. The University of Illinois researchers argue that the dual verbal-numerical presentation has a number of advantages including that of reducing the way respondents interpret ambiguous phrases such as 'very likely' according to their ideology and views about climate change. They are not alone in arguing that verbal probability labels should be accompanied by actual numbers. Researchers with the Union of Concerned Scientists (UCS) in the USA argue that such an approach may reduce the danger that a very strong IPCC statement such as 'it is *very likely* that human activities are causing climate change' could be interpreted as 'we are *not sure* if human activities are causing climate change' (Ekwurzel, Frumhoff and McCarthy, 2012).[13]

Other critics have pointed out that the IPCC's ranges of (ten) probabilities may be too complicated for a general reader, or that they are confusing (as they overlap), or that they ignore affect (how people feel about what they are reading).[14] There is also a debate as to the wisdom of only assigning probability levels to observations in which the scientists have 'high confidence', on the grounds that the combination of high confidence and high probability could be the most useful guide to policy makers. But, as Professor Myles Allen has pointed out, you can have a statement like 'cumulative emissions of carbon dioxide are the principal determinant of the risk of dangerous anthropogenic climate change', in which scientists have very high confidence but, because it has no probability level assigned to it, it might get downplayed or ignored even though it has clear policy implications.[15] Allen argues that too rigid a system of rules about communicating uncertainty may not be helpful, as effective communication depends on the context in which it is being made (such as which policy question is being addressed).

These debates are unlikely to diminish in importance, in part because some sections of future IPCC reports may contain *larger* uncertainties than in previous reports in some parts, such as the sections on possible regional impacts. As mentioned in Chapter 1, some climate scientists have pointed to the apparent paradox that greater understanding of some aspects of climate science can lead to more, not less, uncertainty. For example, Kenneth Trenberth of the National Center for Atmospheric Research in Boulder, Colorado has argued, 'while our knowledge of certain factors [responsible for climate change] does increase, so does our understanding of factors we previously did not account for or even recognise' (Trenberth, 2010). In a commentary on Trenberth's warning, the British journalist Fred Pearce pointed to a series of factors with which computer models are trying to come to grips, such as the role of clouds and aerosols, and the way warming may 'enhance or destabilise existing natural sinks of carbon dioxide and methane in soils, forests, permafrost and beneath the ocean' (Pearce, 2010). Amongst Pearce's conclusions is that both scientists and the public alike are going to have to get used to greater uncertainty in imagining exactly how climate change will play out.

Although the former US Defense Secretary Donald Rumsfeld received much mirth for his articulation of 'known knowns', 'unknown knowns', and 'unknown unknowns', it may be helpful to paraphrase him in this context: 'the number of known unknowns has risen more than the number of known unknowns that climate scientists have transformed into known knowns'.[16]

Exploiting uncertainty

It is a common complaint by mainstream scientists, including many of those working for the IPCC, that sceptics of different types[17] are guilty of picking out the uncertainties, amplifying them, and then using them as grounds for delaying taking action to combat GHG emissions. Researchers in the USA have argued that these sceptics, who are often well organised and well-funded, exploit the uncertainties with the simple aim of spreading doubt (Oreskes and Conway, 2010). The aims and methods of some of the sceptics have been compared to those of scientists and others linked to the tobacco industry who for many years sought to cast doubts on the links between smoking and lung cancer. The media can often act as the main vehicle for amplifying the doubts, particularly in much of the English-speaking press (although not *The Times of India*, the world's largest English-language newspaper). A recent study of 12 newspapers in Brazil, China, France, India, the UK, and the USA strongly suggested that France and the three developing countries stood out for giving considerably less space to sceptics compared to the USA and the UK (Painter and Ashe, 2012). The US newspapers had, by some margin, the highest proportion of articles containing sceptical voices in the periods examined in 2007 and 2009–10.

Well-known sceptics from the UK and other parts of the world are frequently quoted or mentioned in the UK press in a way that specifically links doubts about aspects of climate science to their opposition to taking action to curb GHGs (Painter, 2011: chapter 6 and appendix 4). A relatively limited number of sceptical voices are repeatedly quoted, in contrast to the large number of scientists who agree with the basic tenets of global warming. By far the most quoted or mentioned is Nigel Lawson, the chairman of the London-based Global Warming Policy Foundation (GWPF), but others from Australia and the USA are often present, such as Ian Plimer, a professor of mining geology in Australia and the Republican senator in the USA, James Inhofe.[18]

Sceptics such as these regularly use some version of the view that 'the science is not settled'. What is of particular note for this study is their frequent use of the words 'uncertain' or 'uncertainty' when combined with a message of not taking action. For example, Nigel Lawson's attacks in 2011 on what he called the UK's 'absurd' Climate Change Act were in part based on portraying the impacts from global warming as being full of uncertainties. He was quoted as saying that it was '*highly uncertain* [emphasis added] that higher carbon emissions will warm the planet to a dangerous extent', and added the warning that it was 'futile folly for Britain to act alone when its emissions are two per cent of the global total'.[19]

For his part, Professor Plimer often stresses a different aspect of what he says are the uncertainties surrounding climate science as a reason to oppose a carbon tax. 'There's a huge body of evidence showing no correlation between carbon dioxide and global warming,' he has been quoted as saying (Bevege, 2008). 'Through the geological record we can look back in time and show the Earth has had massive changes in temperature unrelated to carbon dioxide. Why make massive economic decisions when the science is *extraordinarily uncertain*?'

And in the USA, Senator Inhofe has jumped on the language of uncertainty in scientific reports to oppose policies such as cap and trade. In a speech to the Senate in 2005, he said that 'statements made by the National Academy of Sciences (NAS) cannot possibly be considered unequivocal affirmations that man-made global warming is a threat' (Inhofe, 2005). As evidence, he quoted the NAS 2001 report, which used such phrases as 'considerable *uncertainty* in current understanding', 'estimates should be regarded as tentative and subject to future adjustments', 'because of the large and still *uncertain* level of natural variability', and '*uncertainties* in the time histories of various forcing agents' (CRED, 2009).

It is worth stressing that whereas climate sceptics are accused of exaggerating the uncertainties or downplaying the risk of societal impacts, environmental campaign groups are sometimes accused of downplaying uncertainties and overplaying the risks in order to get a more powerful message across. In particular, some climate scientists have criticised them for exaggerating the links between global warming and extreme weather events. For example, in October 2011 Professor Myles Allen wrote in the *Guardian* a rebuttal of Al Gore, in which Gore had been reported as saying that scientists had clear proof that climate change is directly responsible for the extreme and devastating floods, storms, and droughts that displaced millions of people in 2011 (Allen, 2011). Allen argues that it is misleading to think that every time an extreme weather event happens, it's been made more likely by climate change: for some events it will have been made more likely, in others less likely, and in others not much difference.

A final important aspect of uncertainty is the successful promulgation by sceptics of the view that there is significant disagreement amongst scientists. For example, a study carried out in the UK in 2012 by Emily Shuckburgh and others found that 'there was seen to be a lack of consensus in scientific opinion, partly because this is frequently the way climate science is presented in the media. Many focus group participants thought that scientists are often in disagreement with each other or change their mind over time' (Shuckburgh, Robison and Pidgeon, 2012: 19). Work carried out by communications scholar Anthony Leiserowitz and others in the USA suggest that uncertainty generated – deliberately or not – in people's minds about the level of agreement amongst scientists can make a big difference. They describe it as a 'gateway issue' controlling whether or not people choose to engage with the subject of climate change. In other words, people who (wrongly) believe that scientists disagree on global warming tend to feel less certain that global warming is occurring, and show less support for climate policy (Ding et al, 2011). Likewise, research done by the Glasgow Media Group found that in their focus groups, 'the uncertainty of the science left the evidence open to interpretation by a range of experts. Participants said they were very reliant on these interpretations due to the complexity of the subject, and yet found it difficult to place faith in any of them – hence stalemate, and disengagement.'[20]

It's clearly a very difficult challenge for scientists to communicate uncertainties, but the way they talk about them, and the way the media report this, can often act as an obstacle to better public understanding or more engagement. As two experts in climate communication observe, 'scientists do not normally repeat facts that are widely accepted among them, focusing instead on the uncertainties that pose the most challenging problems. As a result, lay observers can get an exaggerated sense of scientific uncertainty and controversy, unless a special effort is made to remind them of the broad areas of scientific agreement' (Pidgeon and Fischhoff, 2011: 36). This is a lesson learnt by Peter Stott, a climate scientist at the UK Met Office, who has had considerable experience of dealing with the UK media. The weight of his experience suggests that 'in dealing with the media, scientists need the space to give the areas they are strongly confident of (even if this is simple stuff), and then the areas where there is complexity.'[21]

3

The Language of Risk

Faced with the difficulties of communicating the uncertainties around climate science, many experts advocate the re-framing of the climate challenge in terms of risk. In this chapter, we first show how risk is part of our everyday experience, and particularly in the insurance, investment, and health sectors. We then outline the arguments in favour of using risk language and metaphors for the climate challenge, and assess its effectiveness in helping policy makers to take action and in helping the general public to understand and engage with the issue. We also include what experienced journalists think about the advantages and disadvantages of using risk language.

The risk society

A close link between risk and environmental issues has a good pedigree in the academic world. In 1986 the German sociologist Ulrich Beck famously described the state of the world as a risk society, which was fundamentally linked to the stage of modernisation (late modernity) that civilisation had reached. Much of the risk he argued was environmental in that risks were mostly related to radioactivity or to the toxins and pollutants in the air, the water or foodstuffs. He said that the large-scale risks of late modernity are different to previous ones in three important respects: they are not limited in time and space; they can be global and catastrophic in their consequences; and they are often not tied to their source or able to be perceived by human senses.[1]

Lord Anthony Giddens, now best known in the UK for his work on the 'third way' in British politics, was famous in the 1990s as the other main author who developed the idea that we now live in a risk society.

Giddens argued, like Beck, that whereas humans have always been vulnerable to 'external' risks like natural disasters, modern societies were subject to 'manufactured' risks such as pollution that were a product of the modernisation process, and particularly a result of innovations in science and technology. Twenty years on, Giddens still argues that we are facing huge and unpredictable risks, particularly from climate change, which humans have never faced before. But he stresses that we can do things to mitigate the risks partly because we live in a world of opportunities too, mostly driven by scientific innovation.[2] He says that he and Beck should have called their books or concept the 'high risk, high opportunity' society, as this captures a great deal of what the twenty-first century will be like.

Both Beck and Giddens are using the word 'risk' in a broad brush sense as the possibility of an adverse impact in the future, but most specialists in risk, like economists, actuaries or statisticians use it in a narrow sense of involving a numerical probability. One of the problems with risk is that it is a slippery concept, meaning different things to different people. For example, for the general public, even the use of the word 'risk' can often convey the sense of a practically negligible 'low probability event'. The phrase 'there's a risk it might rain tomorrow' could be interpreted as meaning that there is a low chance of it raining, even though no numerical probabilities have been assigned to it (CRED, 2009: section 5). And the sort of risks that people have to deal with can vary from the mundane like 'will I miss my train if I stay and have another drink?' to the serious risks associated with a major operation to cure an illness.

However, what is certain is that risk is a common part of everyday activities, many sectors have to deal with it, and people manage it in different ways. For government officials and policy makers, risk assessment is a frequent challenge. Recent examples are the 2011 Fukushima nuclear accident (what were the risks for the population living within a certain radius of the site?), the Icelandic volcanic ash cloud grounding flights in 2010 (what were the risks of a plane crashing?), and the debate about fracking – the process of injecting a mixture of sand, water, and chemicals underground in order to release shale gas (what are the risks of it causing an increase in earthquakes and tremors or environmental harm?).

But for the general public too, risk and uncertainty are part of everyday decision making. The most obvious example is the (mostly) leisure activity of betting on horses or the outcome of sporting events, where individuals are faced with odds about future outcomes and take risks with their money in the face of uncertainty. In 2012, 73 per cent of the UK population over 18 said they had gambled in the last four weeks, although the vast majority of this was via the purchase of National Lottery tickets or scratch cards. The odds of winning the National Lottery are always the same, and so do not require any conscious engagement with the odds, whereas other forms of gambling involve assessing different probabilities. More than 6 per cent of those surveyed (which would mean three million people if scaled up to a percentage of the adult population) had betted on horses or on other sporting events, for which it is fair to assume they had to evaluate probabilities.

Decision making in a context of uncertainty is the essential background to choosing whether to invest in certain stocks and shares, or whether to take an umbrella having seen or read a weather forecast that it might rain. The heated debate about how best to present future information about the weather (deterministic or probabilistic) and the different ways this is done across the world will be taken up in Chapter 4. Insurance is the most obvious example of ordinary people taking out protection against a possible disaster or negative outcome in a context of uncertainty as to why, when, or how something might happen. So they insure not just their cars and homes, but also their bodies and lives against accidents, ill health, or death.

In most Western societies most people opt to have house insurance. For example in the UK, even though it is not a legal requirement (unlike car insurance), there are 26 million homes, of which most are insured against the possibility of fire. There are around 59,000 house fires every year, which works out that there is a 99.78 per cent chance of a fire not occurring in your home. This would be a classic case of the precautionary principle where millions of people take out insurance against low probability, high impact events known as tail risks. British people pay on average £180 a year on building cover insurance (not including content insurance) because they regard the impact of a possible fire as sufficiently worrying to take out insurance. As many mainstream scientists and economists like Lord Stern argue, the climate change risk is greater in size and probability than anything we normally insure against. And some of the risk assessments that ordinary people make are on the same timescale as climate impacts – for example, taking out a pension policy which you will pay into for 40 years is the sort of period over which we will see some of the impacts from climate change.

It is not just the insurance sector where risk assessment is embedded. It is also an essential part of the way most companies make decisions about the future. Indeed, some authors maintain it has been at the heart of economic progress. As Peter Bernstein argues in his best-selling book *Against the Gods: The Remarkable Story of Risk*, companies like Microsoft, Merck, and McDonald's might never have come into being without managing risk. It is their 'capacity to manage risk, and with it the appetite to take risk and make forward-looking choices, [which] are key elements of the energy that drives the economic system forward' (Bernstein, 1996: 3). At a more specific level, many UK companies compile what are known as risk registers when making decisions on certain projects or investments. Using software or other tools, they identify each risk, the probability of an event occurring, its impact, a risk score (probability times impact), counter measures in the case of it happening or preventative measures to reduce the likelihood of it happening. And in the area of health and safety, UK companies are legally obliged to carry out risk assessments.

The military is another sector where assessing risk in the context of uncertainty is an everyday practice. Risk management has been applied for decades by armies around the world. They have to make decisions about global security with incomplete information on a range of threats from international terrorism to nuclear proliferation. The military have to factor in events which have an unknown probability, but a high impact outcome if they were to occur (see for example Briggs, 2012). Indeed, some environmental groups argue that adopting the military's risk management process is a helpful way of approaching climate change risks. As the environmental NGO E3G wrote in a 2011 report, 'risk management [...] is a methodology the national security community has long used when decisions must be made, but information about threats is incomplete, and the future is uncertain' (Mabey and Silverthorne, 2011).

It can be no coincidence that the insurance industry, parts of the investment community, and the military are at the forefront of discussing, or planning for, the risks from climate change, partly because they are dealing with risk every day. The giant German reinsurance company, Munich Re, produces regular and detailed reports about the financial costs of climate impacts, as does the French global insurance company AXA (although their critics would say they have a vested interest in exaggerating the risks).[3] The 2012 report by the US National Intelligence Council is just one of a series since the mid-2000s by the US military and intelligence community stressing that climate change will act as a 'risk multiplier' that has the potential to increase instability in many regions of the world and lead to growing tension over resources and land (National Intelligence Council, 2012).

At a more individual level, making decisions about one's health often involves risk assessment. The health sector, at least in the UK, has invested a vast amount of effort into improving the way risk is communicated to patients. Doctors have to think about risk mainly because of the difficulty of balancing the risks and benefits of any particular treatment for a particular patient. A patient often has to make a decision based on assessing the relative merits of living with the symptoms and the risk of side-effects of the treatment. The NHS and other organisations like NICE (the National Institute for Clinical Excellence) have elaborate guidelines for effective communication about risk. As a senior executive of NICE explains it, these include:

- personalise the risks and benefits as far as possible;
- use absolute risk rather than relative risk (for example, the risk of an event increases from 1 in 1,000 to 2 in 1,000, rather than the risk of the event doubles);
- use natural frequency (for example, ten in 100) rather than a percentage (10 per cent);
- be consistent in the use of data (for example, use the same denominator when comparing risk: 7 in 100 for one risk and 20 in 100 for another, rather than 1 in 14 and 1 in 5);
- present a risk over a defined period of time (months or years) if appropriate (for example, if 100 people are treated for one year, ten will experience a given side effect);
- include both positive and negative framing (for example, treatment will be successful for 97 out of 100 patients and unsuccessful for three out of 100 patients);
- be aware that different people interpret terms such as rare, unusual, and common in different ways, and use numerical data if available. (Jarrett, 2012)

In the UK, local doctors in their health centres promote a website which asks for information about an individual as a way of assessing their chances of a heart attack or stroke in the next ten years, and then provides this information in a variety of numerical and pictorial formats.[4]

The NICE list of recommendations prompts a series of challenges around risk communication, some of which are germane to the discussion of 'effective' communication about climate change. Several studies have shown that making the possible risks of climate change relevant to people's lives promotes more engagement (Glasgow Media Group, 2012; Moser and Dilling, 2004; Shuckburgh, Robison and Pidgeon, 2012: 23), while the interpretation of words such as 'likely' and 'unlikely' used to present possible climate outcomes have also been shown to be problematic. There may be a lot to learn from the way doctors are asked to present the numbers – expressing probabilities as natural frequencies rather than percentages may be helpful to any communication of risk. But the comparisons should not be pushed too far as most health decisions are taken at the individual level, often involve little or no financial sacrifice (at least in the UK), and the outcomes are relatively immediate. In contrast, the worst climate change impacts are probably distant in time and space, most people in the West have no direct experience of them, and the gratification of doing something about them is minimal. And there is little parallel in the health world for the need for collective action to reduce the risks of low probability, high impact weather events that could affect many millions of people some time in the future.

The communication of health risks is cited here more as another example of how many people are making risk-based decisions all the time. But even in the health world, most experts accept that risk is difficult to communicate and there is a lot more to learn and test. They also say the quality of the outcome is very much a factor of patient involvement in a shared decision making process with the clinician about the available options (known as 'no decision about me without me' or 'informed choices'), which has drawn parallels with the need for a 'dialogue approach' to effective climate change communication with the public, rather than the top-down 'information deficit' approach often promoted by governments and scientists (Hulme, 2009a: chapter 7). It is also interesting to note too that at least in the world of health, insurance, and investment, the public often have professional intermediaries who are helping people not only to understand risk, but also to contain or manage it. How you communicate effectively with what groups, with what messages, with which messengers, and with what aims remain a huge set of challenges for whatever sector you are in. But those who advocate strong policy responses to the climate change risk frequently and loudly point out a glaring truism: that in many areas of life, uncertainty about aspects of a future problem is rarely a barrier to action.

Communicating climate risks

The difficulty of portraying and communicating the uncertainties of climate science is one reason why several scientists and communication specialists recommend framing the climate challenge as 'risk in the context of uncertainty'. As already mentioned, it is becoming much more common for climate science reports to speak in this way, particularly when they are aimed at decision makers. The concept and language of risk now often appear in titles of reports, or near the beginning. The UK government's Climate Change Risk Assessment Report of January 2012 is an example of this.[5] In the USA, the 2013 draft Climate Assessment Report in its second paragraph says that 'using scientific information to prepare for these changes in advance provides economic opportunities, and proactively managing the risks will reduce costs over time'.[6] The Australian Climate Commission's 2011 report entitled *The Critical Decade* is another example.[7] In its introduction it lays out the certainties, the near certainties, and the risks. The 2013–14 IPCC reports for the first time will contain a section on 'risk management and the framing of a response'. In the guidelines to authors, the IPCC is also keen to stress the importance of including discussion of 'tail risks', which are low probability, high impact events (Mastrandrea et al, 2012: 1).

Professor Nick Pidgeon of Cardiff University in the UK is one of the most vocal academics advocating the explicit re-framing of the communication of climate change. He wants to focus more on the risks to society rather than the uncertainties within the context of decision making (Pidgeon and Fischhoff, 2011). At least for policy makers, it is clear that the treatment of the climate challenge as risk has some advantages in calculating possible outcomes. As Professor Chris Rapley explains it, one of them is that when insurers measure risk, they weigh up probabilities and impact.[8] When climate scientists emphasise the top of the range of possible outcomes, they can be accused of unjustified bias towards alarmism. But if they are risk assessors, the impact of the top of the range could be huge, so they are obliged to put more emphasis on the upper end, rather than the lower end. In contrast, if you are looking just at uncertainty ranges, the two ends are equal and often excluded in favour of a mid-range 'most likely' figure.

Christopher Hope from the Judge Business School at Cambridge University has shown the importance of probability distributions, as you can end up with very different calculations of the possible economic impact of global warming depending on how you approach uncertainty (Hope, 2011). In his first calculation, he worked out what he regarded as the main economic impacts of the 'most likely' mid-range estimate of global warming of 3.0°C. This was roughly US$200 trillion, an amount less than the cost of taking aggressive mitigation measures to cut GHGs. In other words, a decision maker could argue that, on the basis of this calculation, it would be wrong to take aggressive action as it would cost more to mitigate than it would to suffer the impacts. However, when he made a second calculation, in which he included a probability distribution of a range of inputs to estimate the impacts of temperature rises, then he came up with a much larger mean figure of US$400 trillion because it included the high temperature, high impact possibilities at the top end of the range. A decision maker would make a very different decision based on this figure. Some experts question Hope's first calculation as it does not include the cost of migration from affected areas or the possibility of conflict for example,[9] but the key point here is that ranges of probabilities attached to different possible outcomes can be a better framework for making decisions than a single deterministic or 'most likely' forecast.

The author of the 2006 Stern Report, Lord Nick Stern, is another strong advocate of risk language. He often uses the Russian roulette metaphor to illustrate the high risks we are taking with the climate (see for example Hartcher, 2007). In general, he uses three key arguments:[10]

- The first is that it is what he calls 'the intellectually most honest way' of portraying the climate change challenge. It is not known precisely how much GHG countries will emit, how the temperature will respond (climate sensitivity), and how much impact these temperature increases will have, so at every stage of this chain there are uncertainties. But the risks are centred on what he calls 'some very dangerous places' with potentially some very nasty consequences (particularly at a 5–6°C increase, which he sees as a 50–50 chance if GHG concentrations in the atmosphere reach 750 ppm). He says talking of risks puts to the margins of the debate the idea that somehow these outcomes are remote possibilities or just tail risks.
- Secondly, Stern argues that the uncertainty around climate change is *more* of a reason to worry and take action, not less. He says sceptics who favour a 'wait and see approach' could be seen as reckless when faced with such possible high impact risks, which no wise company

or government would ignore. Sceptics usually argue that they cannot be sure what the impacts will be. But they would have to argue that they *know* that such risks are not going to happen or are very small to make the case that action is not necessary.

- Finally, Stern argues that if you frame the challenge as risk, you can show there is a lot you can do to reduce the risks drastically by lowering GHG emissions. He says you have to break the link between energy and emissions, but the solutions around alternative energy sources have their attractions: 'safer, cleaner, more bio-diverse, and full of innovation and discovery'.

Another argument used against the sceptics is that they often use uncertainty to present the risks in a misleading dichotomy of two alternatives: either the risks exist or they don't, so it is better to wait and see if they really do. Their opponents say this is misleading because risks really exist along a gradual scale of probabilities, of which zero is only one extreme end, rather than one half of the spectrum. In addition, they say sceptics ignore the fact that because GHGs are increasing the risks are growing, and some impacts may be irreversible – waiting and seeing would only be sensible if the risks were not growing and the impacts were definitely reversible.

The use of risk language is often recommended in focus group studies as having clear advantages over the uncertainty framing for public understanding or action. For example, one of the main conclusions of the Shuckburgh study is that re-phrasing statements of scientific uncertainty using the everyday public language of risk might help the problem of their being misunderstood (Shuckburgh, Robison and Pidgeon, 2012: 39). Some argue that the different ways the probabilities involved in climate change predictions are framed can moderate the tendency for uncertainty to undermine individual action. For example, research by psychologists at the University of Essex with focus groups in the UK suggested that it made a difference as to whether a negative or positive frame was used (Morton et al, 2011). The results showed that higher uncertainty combined with a negative frame (highlighting possible losses) decreased individual intentions to behave environmentally. However, when higher uncertainty was combined with a positive frame (highlighting the possibility of losses not materialising) this produced stronger intentions to act.

But the possible advantages of the risk metaphor and language are not clear-cut. Some scientists, like David Stainforth at the LSE, see a need for caution in its use. He argues that a clear distinction should be made between the use of the metaphor in the debate about the *need* for action (where the metaphor can be misleading) and its use in the debate about the *degree* of action that is needed or about what the impacts will be and what should be done to adapt (where the metaphor is helpful). This is how he explains it:

> *Consider the risk of crossing the road or climbing a mountain. These are risks which may be, to some limited extent, quantified. But individuals choose to carry out the activity anyway or, in other words, choose to accept the risk. If we choose to accept the risk, then it may go wrong, but if it doesn't, then we should be fine. If we frame the climate change mitigation challenge as risk, then there is a danger we are misrepresenting it as one of whether we act or don't act. If we don't act, there is 'a risk' that we incur dangerous impacts. If we do act, then everything will be okay. This, of course, doesn't reflect the situation.*
>
> *This is because there are no significant uncertainties that, if mankind continues emitting GHGs to the degree that we currently are, then the impacts for global society will be extremely severe. However, there are uncertainties, which can be discussed in terms of risk, regarding the consequences for any specific location of any specific global response. There are uncertainties too, which can be discussed in terms of risk, regarding the global temperature change resulting from any specific global response strategy. So using risk language when discussing what actions to take is helpful, but not when discussing whether to take action.*[11]

There can be problems with the language of risk as well in terms of how the public understand it. Take for example the case of the phrase 'loading the dice'. This is a metaphor that has often been recommended to climate scientists to express how the build-up of GHGs is increasing the chances of our causing severe disruption to weather patterns, such as more intense flooding (Hassol, 2008). Professor Gavin Schmidt of the Goddard Institute for Space Studies and Professor Myles Allen of Oxford University use the metaphor frequently (see for example quotes in Carrington, 2011; Radford, 2013). However, the Shuckburgh study found that in their focus groups in the UK, 'many simply didn't understand the expression, others couldn't work out why dice were appearing in a report about climate change and some thought something entirely different was being said' (Shuckburgh, Robison and Pidgeon, 2012: 27). Indeed, one member of a focus group thought that 'loading the dice' implied that scientists were somehow fixing the results to get what they wanted or, in other words, cheating. However, it may be that this metaphor works better in the USA where the game of craps in casinos is better known, and thus the concept of loading the dice in that context is understood. In the UK, the game is relatively unknown. This example clearly shows that metaphors have to be constantly tested with different groups to measure their efficacy.

Another related problem is that the language of risk management or assessment can often slip into that of the 'disaster', 'alarmist' or 'catastrophe' language. For this reason, in our content analysis in Chapter 5, a clear distinction is made between the adverse impacts or implicit risks associated with the disaster frame, and the 'explicit' risk frame of risk management and risk language. The type of disaster narrative which is fear-based has frequently been shown to be a good way of attracting media attention but a bad way of improving public understanding, engagement, and behaviour change, particularly when not accompanied by positive messages or concrete examples of what can be done (Moser and Dilling, 2007: 64–80; O'Neill and Nicholson-Cole, 2009; Whitmarsh, 2011). As two researchers from India and Norway express it, 'the more neutral language of risk assessment or risk management [has] not always been successful in shifting positions away from seeing disasters as unavoidable, uncontrollable, unexpected and unprecedented [...] often supported by climate catastrophism' (Khan and Kelman, 2012). Some psychologists argue that risk language can, in certain circumstances, have the effect of creating fear, despair, and anxiety in the mind of the viewer or reader in a similar fashion to the disaster language (Lertzman, 2012).

So, in summary, we can say that the concept or language of risk may be a more effective way of communicating climate change than strong messages of uncertainty or disaster, but it is obviously not a panacea. Risk language may well have clear advantages for officials involved in decision making, but whether it works for the general public's understanding, engagement or behaviour change can vary from audience to audience. As the example of 'loading the dice' shows, even a widely-recommended form of effective communication can – with some audiences – turn out to be an obstacle to better understanding. No matter how plausible it sounds, and how well it might work for some groups or individuals, a particular phrase or frame may not work for others. More research and testing would have to be done with different focus groups to assess its merits. This is because, as a considerable body of research shows, the way individuals or groups of people respond to climate change messages is extraordinarily complex, depending on all sorts of variables like cultural, societal, and political values, personal experience, and the practice or desire for individuals to cohere with values characteristic of groups with which they identify.

Journalists on risk and uncertainty

Finally, we turn to the views of experienced journalists working on environment stories. The first point to emphasise is that the journalists interviewed from many of the newspapers included in our sample in Chapter 5 did not see the reporting of uncertainty as a major problem, although they were aware of some of the difficulties both for them and their readers. They mentioned a lack of understanding from some readers who have a binary vision of 'knowing' versus 'not knowing'; the problem of sceptics jumping on ranges in projections to suggest mainstream scientists do not really know; the pressure from some headline writers or editors wanting to change words like 'might' to 'will', in headlines such as 'climate change "will" kill 100 million people by 2030' rather than 'might'; or the challenge of having enough space to explain where the uncertainties lie.

Their views on risk language were mixed, but generally positive.[12] Most of them also approved of the way the IPCC used the concepts of likelihood and confidence levels (see box). The former *New York Times* environment correspondent and now the author of the Dot Earth blog, Andy Revkin, argues that the house insurance metaphor, for example, has some value:

> It works well in a common sense sort of way – you invest to avoid things you cannot predict well, but don't expect to happen. I spend US$1,600 a year on house insurance, which I do to avoid catastrophe. It's a good way of pointing out that we do face unpredictable hazards, and we do spend money and time to address them. However, the difference is that with global warming, we are not talking so much about our house, but about helping people in Bangladesh minimise sea level rise. So this in the end is a question about values, more than a risk calculation. How much do you value your community, how much do you value avoiding bad outcomes to other people you don't know?

Revkin has also written that the 'insurance against risk' metaphor is not perfect as it does not capture who is likely to pay the biggest premiums to contain global warming, or how you put a financial value on such things as species loss (Revkin, 2010). Alister Doyle, the environment correspondent for Thomson Reuters, believes that, as people understand risk better than uncertainty, it's a help if scientific reports go that way too. He uses the example of the phrase 'there's a 5 per cent risk that global warming will shut down the Gulf Stream', which is easier to grasp when talking about something bad than 'there's a 95 per cent likelihood that the Gulf Stream will be unaffected by climate change'. As he points out, when selling contracts for home insurance the insurer does not say that there's a 99.9 per cent chance that your house won't burn down this year. 'People need to consider even tiny risks with damaging consequences – a category climate change falls into,' he says.

In Australia, the ABC's Sarah Clarke says that translating scientific conclusions in terms of risks can be helpful and mentions doing this in terms of financial risk and through visual representations. On the other hand, the *Sydney Morning Herald*'s Ben Cubby feels 'risk' is often too imprecise a word to be useful in reporting on climate science. From Norway, Guro Tarjem, a reporter for Ekko, a science programme on NRK Radio, believes that 'the risk concept does a better job of communicating the science to the listener than "uncertainty", but uncertainty cannot be ignored. And "risk" is better than uncertainty at creating images in the listener's mind.' Fiona Harvey, formerly of the *Financial Times* and now the *Guardian*, says no metaphor is perfect but the comparison with insurance is helpful as people are familiar with it; in taking out a policy, you have to act as if something will happen. However, she is not a strong advocate of risk language in general as it is often inadequate and can be easily misunderstood.

In France, *Le Monde*'s Stéphane Foucart believes the risk framework is extremely pertinent to addressing the question of climate change. He has made extensive use of statistics and probabilities in his articles to explain risks, which he does not think are an obstacle to the audience's understanding. *Le Figaro*'s Marielle Court, on the other hand, argues that the use of probabilities or ranges in projection found in scientific studies can make the journalist's work particularly hard due to the lack of space to convey the complexity and to translate scientific results into lay people's terms. She also compares the risks surrounding tobacco and climate change:

The correlation between smoking and harm to health is direct and informed. In the case of climate change, it is much more diffuse and with less direct, obvious impact, at least for the population in countries like France. It is therefore difficult to convey these risks, and difficult also for people to get a good understanding of what these risks mean. Indeed they often imply 'you are partly responsible for what's happening'. This is obviously a discourse that people don't want to hear. This, then, leads to debates in most newsrooms, which can lead to conflictive situations among an editorial team.

The final word should go to Curtis Brainard of the *Columbia Journalism Review* (*CJR*), who monitors how the media in the USA report climate change. 'You have to be careful with metaphors. You have to keep in mind the various ways that people might interpret them, and the fact metaphors based on gambling, sports or anything else may work for some groups, but not others. Metaphors can be very effective, but they should be used sparingly. As others have said, "The price of metaphors is eternal vigilance".

Journalists, numbers, and the IPCC

Sarah Clarke, ABC Australia

'In most or all of my stories I would explain the level of confidence scientists might have in a forecast or prediction that they're publishing. However, how much detail or depth I provide depends on the story – TV or radio – and the duration of the piece. For example, if you take the case of the IPCC now being "virtually certain" that humans are causing climate change, as opposed to stating the probability being "99% certainty". I think the "virtually certain" explanation leaves the public open to interpretation. But by quantifying the level of certainty, it leaves the journalist and the scientist under less risk of confusing an issue.

In the past, I have found it difficult to get in touch with the IPCC, or keep abreast of their revisions and updates between the major papers. These days, I've found them more accessible and more open to explanation. The benefits of this are obvious.'

Ben Cubby, Sydney Morning Herald

'If we're talking specifically about the IPCC's use of likelihood words then we'd define them, such as "likely", "very likely", or "virtually certain". You know those words are shorthand for a probability number. So I'd usually explain that in an article if it's relevant. It depends on the context of the piece. Most people know what "likely", "very likely" and "virtually certain" mean in general discourse, and if they're meant to mean something very precise and specific in terms of a probability then we'd usually make that explicit in the story.'

Alister Doyle, environment correspondent for Thomson Reuters

'I do use the IPCC likelihoods in reports such as "it is very likely, or more than 90% probable, that mankind is the main cause of recent warming". It is a lot harder to filter out when a lot of the interesting findings are called "medium confidence", "low confidence" and so on.'

Stéphane Foucart, Le Monde

'Generally, I do not really rely on the IPCC concepts of likelihood and confidence for the nuances they imply are quite often too difficult to grasp for the ordinary reader. However, the IPCC press releases are useful and manage to explain quite well both uncertainties and risks.'

Fiona Harvey, the Guardian

'The scientists have to use the IPCC concepts, and yes, they do help. Scientists have to hedge, as there is nothing we know for certain. However, these concepts can be twisted by unscrupulous people. Scrupulous journalists will report it as 90% certain, but other people will immediately seize upon the 10%, suggesting that the scientists don't know.'

Meena Menon, The Hindu

'We are guided by the IPCC concepts. If we are writing a piece on the IPCC we would quote its text. That is the newspaper's policy. It is important for us to be accurate. The IPCC are not bad, except for the blunder on the retreat of Himalayan glaciers. They did not respond properly to that when the mistake was identified. It gave the Western press a chance to attack Pachauri [the director of the IPCC].'

Andy Revkin, New York Times

'The IPCC concepts are helpful for policy makers so that they have some grounding in what the climate science means in real terms.'

Guro Tarjem, NRK radio, Norway

'I don't think I'd use terms like "likely", "very likely" or "high confidence" in my reporting. Serious journalists will make note of these terms and try to get familiar with them. However, journalists will never grasp them intuitively like a scientist does. Instead of using "very likely", it is better to say that "there is a 90% chance and this is very serious".'

4

Reporting the Future

In this chapter, we look more closely at what we know about the way the media report uncertainty and risk surrounding climate change. There is now a vast body of literature on the media and climate change in general, not just in the USA, UK, and Australia, but also a growing amount on other countries.[1] There is also a large amount of work on the media's portrayals of risk or uncertainty around a variety of topics, including health reporting (see Ashe, 2013). However, much less has been published on media representations of risk and uncertainty around climate science. We review some of the studies that have been done, and then look in more detail at the difficulties for journalists of reporting numbers in general, and probabilities in particular. We discuss the example from the UK of how television's portrayal of ranges of probabilities in weather forecasting might help the public understanding of risk. This is followed by a brief discussion of how more use of visual aids and 'info-graphics' could aid journalists and scientists in presenting risk and uncertainty.

How the media report risk and uncertainty

Many academics and climate scientists take a dim view of the media's portrayal of uncertainty in climate science. Not untypical are comments made by a leading Australian climate professor, Will Steffen, from the Australian National University in Canberra, who has acted as an advisor to different governments. As he expresses it, 'When the science on climate change is so clear, why is it still portrayed as *uncertain* in the media? There's a big divergence between what is known with a high degree of certainty and what is reported' (cited in Lloyd, 2010). He has called it his 'biggest frustration' that a rational discourse is not taking place in his country's media, and urged journalists to focus on the areas where there is less scientific consensus (such as the speed of sea level rise). Professor Steffen was doubtless frustrated by the amount of space given by parts of the Australian media to climate sceptics. This is a common sentiment amongst mainstream climate scientists in other countries too, and particularly the UK and the USA. But before looking at such allegations of 'false balance', we turn briefly to look at media portrayals of risk and uncertainty in other sectors.

A useful overview of the literature can be found in the RISJ 2013 publication 'How the media report risk and uncertainty around science', which includes sections on the drivers of media coverage, and the challenges for scientists and journalists (Ashe, 2013). The broad context is that, in general, news stories about science are often inherently specialist, containing ideas and language that are unfamiliar to most of the lay public and general reporters. Risk and uncertainty in particular are difficult concepts. To name but four difficulties, as we saw in Chapters 2 and 3, there is a wide variety of definitions across academic disciplines; risk and uncertainty can mean different things to the public and experts; risk can involve numbers; and it is hard to untangle where there are more and less amounts of uncertainty around scientific issues.

Another problem is that many 'risk' stories like climate change, nuclear proliferation, population growth or the spread of diseases are slow-burn, creeping issues that do not easily fit the value of newsworthiness, or what some academics call 'first order journalistic norms' like personalisation, dramatisation, and novelty (Boykoff and Boykoff, 2007). As the BBC's environment analyst Roger Harrabin says, (the appearance of) novelty, drama, conflict and personality are the main drivers of news coverage, even about some science issues.[2] He has carried out research on health reporting, suggesting a mismatch between the media's excessive interest in some relatively low risk health areas compared to high risk areas: unusual health hazards such as the SARS virus or vCJD attracted far more press interest than long-term public health concerns such as smoking, alcohol, and obesity (Harrabin et al, 2003). Professor Nick Pidgeon says that conflict, human interest, and personalities are indeed elements likely to attract journalists to risk stories, but adds others like questions of blame, alleged secrets, and cover-ups, and when high numbers of people are exposed to a risk.[3]

It is perhaps not surprising then that some academics have highlighted the shortcomings of the media in reporting risk and uncertainty. Others have sprung to the media's defence (see for example Kitzinger, 1999), and some put the blame more on the shortcomings of science press releases (Riesch and Spiegelhalter, 2011). But when journalists are criticised, it is usually for one of the following:

- *Not making it clear where mainstream certainty and uncertainty on scientific issues lie*: a classic example of not reporting the mainstream scientific consensus robustly was the UK media's treatment of the possible link between the MMR vaccine and autism (see Boyce, 2007).
- *Highlighting worst case scenarios rather than a range of risks*: journalists are often not good at including caveats or qualifiers about ranges of outcomes. In his article examining the media reporting of the *E. coli* food poisoning in Europe in mid-2011, Peter Sandman recommended to public officials dealing with health risks that 'because journalists will tend to drop your maybes and on-the-other-hands and quote the most confident-sounding things you say, [...] "explaining" uncertainty isn't enough. You need to "proclaim" uncertainty' (Sandman and Lanard, 2011).
- *Not distinguishing between numerators and denominators (so-called 'denominator neglect')*: a journalist's job is clearly to cover rare and exceptional events, but the danger comes when they do not make clear how rare these events actually are.[4] In the risk context, 'denominator neglect' can be a focus on the few times when someone at risk actually suffers (the numerator), rather than the many times when those at risk go on unscathed (the denominator). An example of denominator neglect would be the *Daily Express* headline mentioned in Chapter 1, 'daily fry-up boosts cancer risk by 20%'. The failure to mention that it was a 20 per cent increase on a very small number would have left the reader confused about the gravity of the risk (but its inclusion would probably have ruined the story).
- *Not distinguishing between 'absolute' and 'relative' risk*: an important aspect of the L'Aquila case in Italy in October 2012 was whether the public should have been better informed, via scientists and the media, of both the absolute and relative risk of a major earthquake happening when assessing whether to take precautions. In general, in reporting low probability, high impact events like earthquakes, many experts argue that it needs to be explained to the public that the relative risk (i.e. the risk compared to an earlier calculation of risk) could have significantly increased, but the absolute risk

could still remain very small.[5] A headline proclaiming 'Chances of earthquake doubled' (relative risk) could produce panic, whilst a headline 'Minimal chance of earthquake' (very low absolute risk) could produce complacency.

In the specific area of reporting uncertainty in climate science, much of the literature has focused on one aspect of this, which is the amount of space given to sceptical voices. The allegation against journalists is one of 'false balance' or 'balance as bias'. Professor Steffen's comments above have been echoed by a long list of other scientists such as Professor Steve Jones in the UK (in his 2011 report on the BBC) and Professor Schneider in the USA.[6] The essence of their critique is that the media give more space to the different types of climate sceptic than the science would justify, suggesting that there is more controversy than actually exists amongst scientists. They stand accused of not providing enough context, often by failing to mention that many of the sceptics' views are marginal within the climate science community, or by accepting the views of a non-scientist on the science.

There is now a voluminous body of work on this area,[7] so it is enough to say here that it is a more complex picture than many critics of the media would maintain. There are huge variations in the volume, manifestations, and drivers of the media's treatment of sceptics, depending on a wide range of variables including the type of media (print, broadcast or online), the type of newspaper (right- or left-leaning, tabloid or broadsheet), the type of article (news versus opinion), the period in question, and the media landscapes in different countries, as well as societal factors such as the presence of sceptical lobby groups or politicians.

What is pretty much established is how media-generated controversy or misrepresentation of the degree or type of uncertainties around climate science can be an important factor in influencing public perceptions. As we saw in Chapter 3, in 2012 two studies were published in the UK about public understanding of the uncertainties around climate science, and the role that the media play in portraying them. One of the main findings of the Shuckburgh research was that 'there is seen to be a lack of consensus in scientific opinion, partly because this is frequently the way climate science is presented in the media. Many focus group participants thought that scientists are often in disagreement with each other or change their mind over time' (Shuckburgh, Robison and Pidgeon, 2012: 19).

The study carried out by the Glasgow Media Group, which also worked extensively with focus groups in the UK, similarly pinpoints the media's role in the 'construction of uncertainty' in the mind of the public. The study's very first conclusion states that 'there is widespread public confusion over climate change which reflects the journalistic construction of the subject as one of uncertainty. Most people have only a vague understanding of the science, and believe it is inconsistent anyway' (Glasgow Media Group, 2012: 2). This study also highlighted the way politicians are often the most quoted source on climate change (a finding corroborated by other studies). As they are one of the least trusted groups in the UK, this has led to more disengagement from the topic. 'There is a widespread culture of cynicism and distrust, which has led to feelings of powerlessness generally', argues the report.

Media coverage can send the message to readers and viewers that the science is uncertain without ever mentioning the word 'uncertainty' in stories. Of particular relevance to this study is research by two Americans who designed an experiment with readers involving three different newspaper story treatments – controversy, context, and control (neither context nor controversy). They found that greater contextualisation within climate science stories helped to soften the controversy stirred up through uncertainty (Corbett and Durfee, 2004). In other words, the inclusion of an assessment where the majority of scientists' views lie in what is known or uncertain can mitigate the effect of a controversy highlighted in the media and prompted by uncertainty. The lack of contextualisation, and in particular the presence of 'duelling experts' without a sense of where the consensus lies, can hinder public understanding.

There have not been many studies which look specifically at the media's more general portrayal of uncertainty or risk around climate science, or which carry out detailed content analysis of just how the media reports them. Of those that have done it, in one of the earliest studies, US academic Stephen Zehr looked at the representation of scientific uncertainty around climate change in the US print media from 1986 to 1995. He concluded that uncertainty was represented in several forms, and that uncertainty 'was used to help construct an exclusionary boundary between the public and climate change scientists', thereby contributing to deferential citizens and diffused public involvement through acceptance of the need for more research (Zehr, 2000: 85).

Other studies have shown the different ways uncertainty has been portrayed in different media within the same country, or between the media in different countries. Bob Ward from the Grantham Research Institute in London has argued that the various media outlets in the UK have adopted distinct approaches, in part driven by ideological differences, in reporting the link between climate changes and weather-related natural disasters, and the uncertainties involved (Ward, 2008). The Swedish academic Ulrika Olausson showed from a study of articles in three Swedish newspapers in 2004–5 that the media – at least in Sweden – were reluctant to display any kind of scientific uncertainty that would undermine the demand for collective action (Olausson, 2009).

Media researchers have spent more effort on analysing the different 'framing' by the media of the climate change narrative. There is a vast literature on framing theory, and the way the media frame particular stories (see for example Entman, 1993; Nisbet and Mooney, 2007). In essence, framing a story means selecting some aspects of a perceived reality and making them more salient than others in a text. In other words, frames tend to privilege some aspects of a complex problem rather than others. This can have the effect of providing a perspective from which a reader or viewer can interpret a problem by stressing some aspects of it or by ignoring or downplaying other aspects, and can deeply influence how persuasive consumers find that information. Frames can be triggered by words (sometimes called linguistic repertoire or discourse), but also by non-verbal or visual prompts, particularly in broadcast media, such as tone of voice, gestures, imagery, music, and who the messengers are (Moser, 2010: 39). All sorts of influences come to bear on how a journalist frames a story, including editorial practice, news criteria, the power of the messenger, or the journalist's conscious or unconscious ideology.

Much work has gone into assessing the dominant or strong frame of climate change stories, and different authors have selected a wide variety of different frames to assess them.[8] Many of their studies show that a variant of the disaster frame is often dominant or frequently present in the headline or the body of the text. These disaster frames share an emphasis on general or specific adverse consequences or impacts from climate change such as more intense hurricanes, species death, ice melting, or sea level flooding, and effects on humans such as population displacements, food shortages or health problems. Sometimes, but not always, this disaster frame is accompanied by the language of catastrophe, which some researchers code separately as a different frame such as 'alarmism', 'catastrophe' or 'fear'.

For example, a study by the British climate scientist Professor Mike Hulme found that in virtually all the UK print media, easily the most common tone of the reporting of the IPCC's 2007 WG-1 and WG-2 Fourth Assessment reports was alarmist, dominated by the language of catastrophe, fear, disaster, and death (Hulme, 2009b: 124). Over 75 per cent of the articles fell into this category. He also concluded that the reporting of WG-1, which was about the climate system, included 'embellished interpretations' of the impacts (not included in the WG-1), which were 'reported by recycling previously published accounts and reports, or through creative imagination'. It can be argued in defence of journalists that the WG-1 report tended to lack context, so a temperature rise, for example, had to be translated into possible impacts in order to have more salience with readers.[9]

In similar fashion, two researchers at the Tyndall Centre at the University of East Anglia looked at 150 articles in the UK quality press from 1997 to 2007, and examined which of five discourses (optimism, rationalism, ethical or self-righteous mitigation, 'disaster strikes' or potential catastrophe, and opportunity) were the most common (Doulton and Brown, 2009). An example of the 'disaster strikes' discourse was 'They're going under: two islands have disappeared beneath the Pacific Ocean – sunk by global warming'; of the potential catastrophe discourse, 'Ten years to prevent catastrophe'; and of the opportunity discourse, 'simple renewable energy technology can be used both to adapt to the threat of climate change and also lift people out of a subsistence existence'. They found that 'potential catastrophe' was by far the most common discourse, accounting for a third of the 150 articles, whilst 'disaster strikes' was also relatively common, with around 20 articles.

In sharp contrast, 'optimism' and 'opportunity' corresponded to only five articles each. The authors were particularly interested in the link between development and climate change, so they were looking for frames which stressed the opportunities arising from the benefits for the developing world to be had from switching to clean energy, and in kick-starting a move towards sustainability. This research supports those who argue that, in general, the media tend to focus more on dramatic representations of climate science (scary, doom-laden scenarios using the language of fear and disaster) than the language of opportunity and solutions. For example, the IPCC's 2007 WG-3 report on mitigation, which included policy options, received significantly less coverage in the UK print media (although this may have been due to competing stories, and its launch in Bangkok compared to Paris for the WG-1 and Brussels for WG-2) (Hulme, 2009b: 123).

Such a prioritisation of frames is seen as very significant by communication analysts. In general alarmist or fear-based messages are more likely to induce apathy or paralysis through powerlessness or disbelief than motivation and engagement – particularly if not accompanied by an action strategy to reduce the perceived risk (Moser and Dilling, 2007: 64–80; O'Neill and Nicholson-Cole, 2009). For example, Professor Mike Hulme says shrill voices crying doom can paralyse instead of inspire. As he has expressed it, 'I have found myself increasingly chastised by climate change campaigners when my public statements and lectures on climate change have not satisfied their thirst for environmental drama. I believe climate change is real, must be faced and action taken. But the discourse of catastrophe is in danger of tipping society onto a negative, depressive and reactionary trajectory' (as quoted in Revkin 2007).

Max Boykoff's study of news articles on climate change in the four main daily tabloid newspapers in the UK (the *Sun*, *Mirror*, *Express*, and *Mail*) from 2000 to 2006 also found that 'headlines with tones of fear, misery and doom were most prevalent' (Boykoff, 2008). Through a combination of content analysis and interviews with journalists, Boykoff concluded that the articles were 'predominantly framed through weather events, charismatic 'mega-fauna' (like polar bears) and the movements of political actors and rhetoric, while few stories focused on climate justice and risk'. By the 'climate justice and risk' frame, he did not mean the narrow sense of risk we have been using in earlier chapters, but a discussion of the ethics surrounding the different vulnerabilities of low-income and better-off sectors of society, and their different abilities to cope with climate change impacts.

The examples above have been taken from the UK and from print media. However, a study in 2012 of three popular US news websites run by broadcast companies (Fox News, MSNBC, and CNN) from 2007 to 2009 showed that 'environmental catastrophe' and 'scientific/technical uncertainty' were two of the top four most common frames in the headlines (along with 'strategy/conflict' and 'public accountability/governance') out of the 12 present (Boenker, 2012). Likewise, a 2006 study of eight Norwegian newspapers found that the main focus in the reporting was that of an imminent climate catastrophe; uncertainties also were given attention, mostly by climate scientists who were quoted relatively often (Ryghaug, 2006).

So, in summary, there is strong evidence from other studies that the media frequently use the disaster and/or uncertainty frames in their climate change stories. The opportunity frame is seldom present, although more specific studies of media coverage of the wide range of reports stressing the green economy or other opportunities from low-carbon development might yield different results. There is also a gap in the literature examining the media's framing of climate change as 'explicit risk', which, in the narrow definition expanded in the next chapter, can include the use of numbers to assign probabilities to the likelihood of adverse outcomes and to calculate their possible consequences.

Strength in numbers

It is a common complaint from media observers that many journalists are not very good with numbers and probabilities, or that they worry too much that their readers or audiences will not understand them. The former head of the Institute of Fiscal Studies in the UK and now warden of Nuffield College at Oxford University, Andrew Dilnot, has given a number of RISJ seminars pointing out how journalists frequently misinterpret or misrepresent numbers, including not understanding means and medians, or writing misleading headlines. In similar fashion, the *Financial Times* journalist and presenter of the BBC radio programme *More or Less*, Tim Harford, argues that journalists like to pepper their stories with numbers because they look like facts, but in fact 'the principle of the good use of statistics is no different to the principles of good journalism: ask if the numbers are true, ask what do they really mean, and ask what's the bigger story or wider context?'[10]

The former director general of the BBC Mark Thompson told an Oxford audience in 2012 that it would be a major advance if more BBC journalists could understand risk and statistics to improve their reporting.[11] One obstacle is that few British journalists have a background in science or maths. The 2013 report by the National Council for the Training of Journalists (NCTJ) in the UK showed that only 5 per cent of the 1,000-plus journalists surveyed had undergraduate degrees in science, compared to 23 per cent in literature, 16 per cent in social studies (including politics and economics) and 12 per cent in history (NCTJ, 2013: 31). The number with a maths degree is likely to be even fewer than science as it was one of several subjects featured in the 10 per cent classified as 'other'. What is more, few journalists get numeracy training at their workplace.

This is a particular problem when the number of specialist journalists, including environment reporters, is declining in the Western media and there are more general reporters dipping into previously specialist areas. Secondly, more and more journalism involves numbers from 'big data' to economics, opinion surveys, crime statistics, ageing populations, and health and environmental risks, to name but a few. In the USA, the *New York Times* blogger and data processor Nate Silver received a huge amount of publicity for correctly predicting the outcome of the 2012 elections in all 50 states. As the science writer Frank Swift argued in a *Guardian* article calling for more science education for journalists, 'the triumph of Nate Silver's data-driven election forecasts bruised the egos of American journalists who'd clung to conventional tools like political pundits and vox-pops, but the implication is clear: embrace the world of data or face irrelevance.'[12]

A strong case can be made that journalists need to be better not just at understanding data, but at understanding and presenting the numbers around risk and uncertainty. Considerable effort has been directed at the better reporting of health risk, with a number of organisations and academics in the UK and USA providing advice and guidelines (see for example LaFountain, 2004; Science Media Centre, 2012). There is some evidence that at least in the UK there have been improvements. The NHS Choices site, which monitors media coverage of health issues, reckons that in 2012, with the exception of the tabloid *Daily Express*, fewer 'wonder cures' were hitting the headlines and peer-reviewed medical reports were covered more responsibly. But, 'journalists still cannot be trusted with the evidence, especially when it is a matter of probability and subtle distinction between absolute and relative risk.'[13]

Health reporters may be better at reporting risk as they are under more scrutiny – their coverage can have a direct and immediate impact on readers' lives. But covering probabilities and risk is a hugely important challenge for climate change reporting too. For example, Hurricane Sandy in October 2012 prompted a lot of media discussion in the USA as to whether, or to what extent, it could be linked to human-made global warming. Scientists differed as to whether the media did a good job, and some were very critical of headlines like Bloomberg Businessweek (1 November 2012) which put on its front page in large letters, 'It's Global Warming, stupid'. But one of the most telling observations came from Professor Tom Knutson, a meteorology expert with the US Geophysical Fluid Dynamics Laboratory, who said,

A trap reporters can fall into is chasing after answers to a poorly worded straw man question: Did climate change cause this event? Instead the science will generally only be able to look at questions of attribution in a probabilistic sense: Has climate change altered the odds of events like this one occurring? (Pidcock, 2012b)

So better reporting involves better portrayal of probabilities. If it is true that journalists find probabilities difficult, then, at least in the UK, they are hardly alone. There is a lot of evidence that the general public also find them hard to grasp. For example, many people believe intuitively that after nine throws of heads, the odds of tails are increased on the tenth throw of a coin.[14] And when a 2012 survey by the Royal Statistical Society's 'getstats' campaign asked MPs to give the probability of getting two heads when tossing a coin twice, more than half failed to get the answer correct (25 per cent or one in four). However, the politicians fared better than the general public. In a similar survey conducted in 2010, just a third of the 1,028 respondents answered correctly.[15]

It is notoriously difficult to communicate probabilities to lay audiences. However, some risk experts have put forward a convincing case that much more could be done, at least in the UK, to familiarise journalists, policy makers, and the general public with probabilities by changing the media's presentation, particularly on television, of short-term and seasonal weather forecasting. In the USA, TV forecasts giving probabilities of precipitation (rain, sleet, and snow) have been provided by the National Weather Service since the 1960s. It is also true of Canada. Weather forecasts on television in these countries are usually accompanied by information giving such probabilities and indicators of ranges (such as how strong the UV content of sun will be).

In the UK, the tradition has been to give what are known as 'deterministic' rather than 'probabilistic' forecasts. A 'deterministic' forecast is a single 'best bet' prediction of what is most likely to happen like 'sunny intervals', 'occasional showers' or 'overcast'; a 'probabilistic' forecast can be providing odds for a range of temperature forecasts, or just one number attached to the probability of precipitation like 'a 30 per cent chance of rain falling in London'. Deterministic forecasts are also the norm on television in the four other countries included in this study, Australia, France, India, and Norway, although, as in the UK, official meteorological websites often include probabilistic forecasts.

One of the advantages of probabilistic forecasts is that the Met Office in the UK, which is the main body giving predictions about the weather, could have avoided the ridicule heaped upon it in the tabloid press after its press office predicted 'a barbecue summer' for 2009 (which turned out to be a washout). Another example is the famous case in October 1987 when a BBC weather presenter, Michael Fish, told viewers that a woman had rung him to say that she was worried about reports of a hurricane coming. He added with a wry smile that 'if you're watching, don't worry, there isn't!' That evening, the worst storm to hit south-east England since 1703 caused record damage and killed 18 people.

Climate scientists now use thousands of daily observations from across the world to design computer models that replicate the climate system, and to make weather forecasts over short and longer periods. They run slightly different versions of the models, and start from different initial weather conditions, to determine the most and least likely weather outcomes, which are called ensemble forecasts. They are never entirely accurate, and the longer the period of forecast ahead, the less accurate they are. Tiny uncertainties at the start of the modelling, which can come from weather observations which are not completely accurate or which do not completely cover the atmosphere, can produce different results. This is what actually happened in October 1987 as a tiny uncertainty over the position of a low pressure system over the Atlantic made an enormous impact of whether it was amplified or not. Professor Tim Palmer of Oxford University has gone back and run the models which in fact show that 30 per cent of them were showing hurricane force winds coming to south-east England.

Professor Palmer is one of many scientists strongly advocating the use of probabilistic forecasts on television, in part because it shows the public that uncertainty does not mean scientists do not know anything. On the contrary he says, uncertainty can often be quantified in helpful ways as a basis for making decisions. Giving ranges of probabilities and reliability or confidence levels are essential. So for example he says 'a good journalist would have asked the Met Office in the case of the 2009 summer forecast: what is the level of probability of a barbecue summer; and how many times have you been right in the past when you have made such a prediction?'[16] Another advantage is that a probabilistic forecast is more likely to include the chances of a low probability, high impact event like a hurricane which may have only a 10 per cent chance of happening but could cause an enormous amount of damage.

Some believe that the main reason why the UK media do not present probabilistic forecasting is the view amongst many editors that the British people cannot understand probabilities.[17] This may be the case but, even if everybody were to be more numerate, it remains an enormous challenge for climate scientists, weather forecasters, and journalists to find the best way – out of all the options – of communicating the risks, uncertainties, and probabilities. There are several pitfalls, some of which are likely to be given voice in parts of the media. One is that presenting more, or too much, information to the public can be confusing. Another is that giving probabilities could attract the accusation that the forecasters don't really know, or that they are merely trying to deflect blame if they get it wrong. An article in the *Daily Mail* in November 2011 responded to the Met Office's new way of presenting forecasts online with the headline 'New Met Office forecast system likely to mean 80% chance of confusion' (Cohen, 2011).

The Met Office ran an online weather game in 2011–12 together with researchers from the Universities of Cambridge and Bristol, which was one of the largest projects of its kind ever carried out.[18] It involved helping an ice-cream vendor named Brad to make decisions over when to sell his products over a four-week period depending on an assessment of the weather uncertainties. Different presentations of uncertainty involving graphics and written probabilities were given. To cut a long story short, the top line result from most of the several thousand taking part in the game was that understanding the probabilities would not be an obstacle to making good risk-based decisions. As Liz Stephens from Bristol University says, 'while educational attainment and age showed some influence, in general the users were not that confused by the uncertainties involved. As long as the probability was provided, the exact method of presentation did not greatly affect this understanding'.[19]

Ken Mylne, the ensemble forecasting manager at the Met Office, says that the move to probabilistic forecasting is 'more likely to be a marathon than a sprint', but the key is public familiarity. 'We need to start getting probabilistic forecasting out there more. Then it will be so much easier for people to understand things like seasonal and long-range forecasts' (Pidcock, 2012a). Clearly there is a role for television, but also for other visual representations on different media platforms of complex numerical information covering risk and uncertainty.[20]

There are now a wide variety of ways for the media to present risk information visually from 'cones of uncertainty' depicting different possible hurricane paths, to risk ladders, icon arrays, probability contours for wind gusts, or tree maps.[21] Crucially, the same risk information can be presented in a combination of formats online to take account of the varied ways different people absorb this type of information or need to use it. There is already a rich collection of academic and more popular articles exploring the visualisation of uncertainty and risk, and the benefits or downsides of different approaches (see for example Spiegelhalter et al, 2011; Stephens et al, 2012). Examples have been examined from the world of sport, weather, climate, health, economics, and politics, but there is still a lot to understand about how different types of visualisation are processed and understood by lay audiences.

There are also a lot more opportunities for mainstream media and online sites to present complex climate science information visually. The technological advances in recent years have massively increased the possibilities for much better visual presentations on newspaper and other media websites, let alone on tablets and other platforms. Richard Black, the BBC website's former environment correspondent, gives the example of the way the BBC website portrayed some of the data from the 2007 IPCC reports. He says one rather clunky chart was 'the funkiest thing we did with the IPCC report, but now it looks way out of touch'.[22] Sarah Clarke, the environment and science correspondent for the ABC in Australia, says that in a January 2013 story examining sea level rise, she presented the risks associated with climate change in the most visual way possible, using maps and areas of high population to spell out the different levels of flooding over time.[23] She thinks 'viewers are better able to comprehend the conclusions if it's translated to "impact" and "risks" and visually demonstrated how it would affect their environment. Graphic representation and pictures are the key to assisting the understanding of climate change science.'[24]

Other journalists and media critics interviewed for this study highlighted other examples of where, in their view, info-graphics clearly help understanding of risks and uncertainty, although many stressed too that the data can be manipulated or falsely presented to serve an agenda. Andy Revkin is a fan of the 'Burning Embers' graphic depicting climate risks, which was actually left out of the 2007 IPCC report.[25] Alister Doyle points to the 'greenhouse gamble' developed by the Massachusetts Institute of Technology which is designed to convey uncertainty in climate change prediction: you spin roulette wheels and end up with anywhere from 3 to 7+ degrees of warming depending on the choice of policies to curb GHG emissions.[26] And *CJR*'s Curtis Brainard is an admirer of the UCS' graphic describing the strength of scientific evidence, from low to high, linking climate change to changes to different extreme weather events, and of the IPCC's chart describing the confidence levels, which he says should be included in more stories about climate change and uncertainty.[27]

Finally, it is interesting to note that some newspapers in the West which are keenly trying to re-invent themselves and find a new business model in the face of declining revenues, are making info-graphics in print, online, and tablet a key element of their re-launch. Juan Señor is a former RISJ fellow and partner of Innovation, a company which is one of the market leaders in advising newspaper owners how to find new formats, content, and ways of working. He says that the technological advances around info-graphics make it an ideal way to present complicated information and also to show a story rather than tell it: 'Innovative, interesting and clear info-graphics have to be a necessary part of the mix of the future offer on different platforms. The presentation of climate risks is perfect for info-graphics.'[28]

5

Uncertainty and Risk in the International Print Media

The preceding chapters suggest a wide range of issues that arise concerning the media's coverage of uncertainty and risk, both in general and around climate science. It is helpful to pick out and develop some key points to provide the essential context for our study:

1   Uncertainty is a common feature of the way the media report science topics, but it is rarely spelt out that uncertainty is an essential element of research science and can be something positive. This uncertainty can appear in several forms, often without ever mentioning the word 'uncertain'. These can include language such as 'may', 'might' or 'could'; representations of controversy such as duelling experts; new research which raises more uncertainties, disagreements, or the need for further study; and ranges of results, including uncertainty parameters, likelihood indicators, and confidence levels.

2   There are a wide range of uncertainties around climate change and in particular the timing, extent, and location of impacts in the future. The level of uncertainty can be amplified by the presence in media treatments of dissenting sceptical, often non-scientific voices, which do not necessarily represent the scientific consensus. A linked problem is the lack of context as to where this consensus lies.

3   The media in several countries often use variations of the 'disaster', 'catastrophe' or 'alarmist' frame when they are covering climate change. The different manifestations of this disaster frame range from a strong version of 'catastrophe' or 'alarmist' at one end of the scale to a softer 'alarming' version stressing adverse or negative effects, impacts or consequences resulting from warmer temperatures.

4   In comparison with the disaster frames, there are many fewer articles about climate change where the opportunities offered by mitigating

the risk of climate change is highlighted or included. For example, there is a dearth of articles which speak of the uncertainties or negative impacts, and at the same time speak of opportunities such as the move to a low-carbon economy and/or renewable energy sources as a way of reducing the risks.

5   Different frames in news articles can be measured or assessed by their presence anywhere in an article, or by their salience (presence in headlines or opening paragraphs), or by their dominance throughout an article (which includes a wide variety of indicators such as the relative weight of a frame throughout an article, salience, prominent quotes, and the use of language such as metaphors and adjectives).

6   Establishing the presence, salience or dominance of frames can give important insights into how audiences might receive and understand them. For example, early or strong messages about uncertainty in articles can be the dominant impression left on the reader or viewer, which can act as a strong incentive for disengagement. Words such as 'may' or 'might' used to signify uncertainty can often be interpreted by lay people to mean that scientists are ignorant rather than uncertain. Research with focus groups also suggests that a strong disaster framing is good at attracting attention, but can often act as an obstacle for public understanding and engagement.

7   In recent years, there has been a growing trend of using the language of risk in the general presentation of the climate change challenge or in specific reports. This often takes the form of using the concept of risk in a narrow or statistical sense where numbers and probabilities are included, or where risk metaphors such as 'Russian roulette', 'taking out insurance' or 'loading the dice' are mentioned. There are few, if any, studies which examine how this narrow definition of risk is included in media treatments and, if so, when and how it is included at the same time as other frames of uncertainty and disaster.

For our study, we wanted to test the presence, salience, and dominance of different frames across different aspects of the climate change story, in different newspapers and in different countries. This wide range of variables was intended to give a robust sense of which frames were particularly present in the presentation of information about climate change across the globe despite the different political and media contexts in which it was presented.

In line with the key points listed above, the *four key frames* we tested were uncertainty, disaster/implicit risk, explicit risk, and opportunity. As regards the two risk frames, a clear distinction was made between:

1 The disaster or 'implicit risk' frame, which included adverse impacts or effects such as sea level rises, more floods, water or food shortages, population displacements, and so on; in the case of Arctic sea ice melt, negative effects on the ocean ecosystem and nations living on the Arctic rim, heightened political tension, and the effect on the weather in high and intermediate latitudes (with possible links to unusual cold snaps in Europe and the USA). We did not distinguish between harder or softer versions of this frame as described above, but included all manifestations of the adverse consequences whether or not they were accompanied by strong fear-based language.
2 'Explicit risk', where the word 'risk' is used, or where the odds, probabilities or chance of something adverse happening are given, or the inclusion of everyday risk concepts or language like insurance or betting.

As regards the opportunity frame, we included two different types of opportunity:

1 those associated with reducing the risks of climate change (moves towards cleaner, safer, more bio-diverse sources of energy or low-carbon development); and
2 those associated with *not* reducing the risks of climate change. In the latter case, in the context of mild global or regional temperature rises, these were examples of where there would be scope for growing more or different agricultural products, a more pleasant climate, fewer deaths from colds, and so on; in the context of Arctic sea ice melt, these were economic opportunities in trade, fishing, oil, gas or mineral exploration/drilling, and/or new shipping routes.

Examples of the four frames can be found in Table 5.1, for headline, dominant tone, quotes, and language. So for the frame 'explicit risk', an example of a headline would be 'Hundreds of millions of extra people at greater risk of food and water shortages'; (partial) evidence of a dominant tone would be 'the entire coastline will face up to 20 per cent increased risk of cyclonic storms'; of a quote 'The basic economics of risk point very strongly to action'; and of language, 'risk', 'insurance', 'precautionary principle', 'loading the dice', or 'playing roulette'.

Secondly, the *different types of climate change stories* we examined were the media's reporting of the IPCC's first two reports of 2007, the IPCC's special report on extreme weather of March 2012 (SREX), and the reporting of the decline of Arctic sea ice in the last two to three years (see box). The IPCC's 2007 reports were chosen because they attracted

Table 5.1   Examples of Indicators of Frames

Frame	Uncertainty	Disaster/ implicit risk	Explicit risk	Opportunity
Headline	'Climate change effects unknown'	'More wild weather on the way, UN climate panel says'	'Hundreds of millions of extra people at greater risk of food and water shortages' 'a third of all species at risk'	'The silver lining to Arctic global warming'
Dominant tone	'Great uncertainty remains about how much of an impact climate change will have on future extreme weather events'	'Billions will be hit by flood and famine'	'The entire coastline will face up to 20% increased risk of cyclonic storms'	'Thaw opens up new shipping route'
Quotes	'Right now the whole debate is polarised'	'By 2020, 75–250 million people will be exposed to water scarcity due to climate change'	'The basic economics of risk point very strongly to action' 'society needs to take a massive insurance against this horrific prospect'	'There will also be geopolitical as well as commercial opportunities'
Language	'fiercely disputed'; 'casting doubt on predictions'	'catastrophic'; 'severe', 'deadly' or 'costly' consequences	'risk, insurance, precautionary principle, loading the dice, roulette'	'huge benefits'

a huge amount of coverage all around the world and were one of the reasons for the peak in the volume of international reporting of climate change around this time.[1] The SREX report attracted much less coverage but was significant for our study because of the explicit inclusion of risk. The Arctic sea ice story was selected because it was widely covered by the international press due to its obvious newsworthiness. The World Meteorological Organization called it a 'clear and alarming sign of climate change'.[2] Clearly, the topics are very different in that the IPCC reports are essentially designed to inform policy, whereas the Arctic story is something happening in the real world. However, the intention was to choose two widely covered but different aspects of the climate change story to examine the presence of the different frames.

Moreover, all four stories included uncertainties, risks, and opportunities. The two IPCC reports of 2007 included the concepts of likelihood and levels of confidence in their findings. Because numerical probabilities were assigned to the likelihoods of particular outcomes, they fitted our criteria for the 'explicit risk' frame. Our analysis therefore included the extent to which these concepts were used and explained in the media's coverage. The IPCC's SREX report combined uncertainty and explicit risk, which made it a useful focus of research to illustrate how the media reported on these two aspects which were at the centre of a major new study on climate change. The Arctic sea ice story combines uncertainties, risks, and opportunities. It represents a certain type of opportunity brought on by not doing anything to reduce GHG emissions. As the ice melts, there will be more interest in oil and gas exploration as once-frozen areas are unlocked for longer periods – the area north of the Arctic Circle is thought to contain 13 per cent of the world's undiscovered oil and 30 per cent of its undiscovered natural gas. In addition, more ships will start using the northern sea routes (Clark, 2013; *Economist*, 2012).

Thirdly, we looked at *three different newspapers in six different countries* (Australia, France, India, Norway, the UK, and the USA). These countries were chosen in order to have a wide range of different media landscapes and journalistic practices, different political, economic, and social contexts, and different media treatments of climate change. These differences are explored in the next chapter, but they include the following: in France and India, climate change is not subject to the same degree of intense political controversy as it is in Australia, the USA, and the UK; climate sceptics are present in the media and in the wider political setting in these three countries and Norway, but not in India and not so much in France; the UK has a strong tabloid press compared to France, India, and the USA; Norway is the only oil exporting nation of the six; India is included as an example of a developing country where the climate change debate is often framed as one of development versus mitigation or climate justice in the international sphere; and there are marked differences between the degree to which the general public in the six countries perceive the importance of climate change or the robustness of the science.

The three newspapers in each country were selected to represent different styles, specialisms, readerships, political sympathies, and ways of reporting climate change in order to achieve a wide spectrum of newspaper profiles in the sample. Again, a full description of each newspaper can be found in the next chapter, but the principal aim was to include in each country a liberal or left-leaning and right-leaning paper, and either a business or tabloid newspaper. So in Australia we included the liberal *Sydney Morning Herald*, the right-leaning the *Australian*, and the tabloid *Herald Sun*; in France, the left-leaning *Le Monde*, the right-leaning *Le Figaro*, and the tabloid *Le Parisien*; in India, the left-leaning *The Hindu*, the more centre-ground *The Times of India*, and the business newspaper the *Business Standard*; in Norway, the traditionally right-leaning paper *Aftenposten*, the less ideological *Verdens Gan*, and the tabloid *Dagbladet*; in the UK, the left-leaning *Guardian*, the right-leaning *Daily Telegraph*, and the tabloid *Daily Mail*; and in the USA, the liberal *New York Times*, the centre-ground *USA Today*, and the business newspaper *Wall Street Journal*.

The four case studies examined

1. The first part of the 2007 Fourth Assessment IPCC Report (known as WG-1) was launched on 2 February. It concentrated on the physical science of climate change, and summarised and assessed the current scientific knowledge of the natural versus human drivers of climate change. Its headline findings were that the 'warming of the climate system is unequivocal', and 'most of the observed increase in global average temperatures since the mid-twentieth century is very likely due to the observed increase in anthropogenic greenhouse gas concentrations'.

2. The second part of the 2007 Fourth Assessment IPCC report (known as WG-2) was launched on 6 April. It concentrated on

impacts, adaptation, and vulnerability. It stated that 'evidence from all continents and most oceans shows that many natural systems are being affected by regional climate changes, particularly temperature increases'. It also described some of what might be expected in the coming century, based on studies and model projections, including changes to fresh water supply, ecosystems, food production, and coastal systems.

Both reports relied on computer modelling and other predictive methods to make projections for ranges of temperature rises and their possible impacts. They were notable for including both levels of likelihood and levels of confidence such as 'very likely' meaning more than 90 per cent probability of something happening, and 'very high confidence' meaning 'at least a nine out of ten chance'. Together with a later report on mitigation (WG-3) and a synthesis report, they were widely regarded as constituting the most comprehensive international assessment of the state of scientific knowledge about climate change at the time. The reports were produced by thousands of authors, editors, and reviewers from dozens of countries, citing over 6,000 peer-reviewed scientific studies.

3. The IPCC's March 2012 report on extreme weather events. Its full title was 'Special Report on Managing the Risks of Extreme Events and Disasters to Advance Climate Change Adaptation', known in short as SREX. The report explored the social as well as physical dimensions of weather- and climate-related disasters, and considered opportunities for managing risks at the local to international scale. The report explicitly included the concept of risk in its title. The press release of 28 March mentioned the word 'risk' nine times, while the report itself included it 4,387 times according to an article in the *The Times of India* published the day after. One of the main coordinators of the report, Chris Field, stressed that 'the main message from the report is that we know enough to make good decisions about *managing the risks* of climate-related disasters. The challenge for the future has one dimension focused on improving the knowledge base and one on empowering good decisions, even for those situations *where there is lots of uncertainty* [emphasis added].'

4. The Arctic sea ice melt since January 2010. On 16 September 2012, the summer ice cover over the Arctic Ocean reached a new record low. Arctic sea ice follows a cycle, growing during the winter and retreating during the summer. The record summer melt of 2012 was not unexpected, as in recent years the ice cover has consistently fallen below the average extent for the period 1979 to 2000. But the fact that it had shattered the previous record in 2007 by some margin provoked a series of press articles around the world, which at times included the reaction of many scientists working in the field as one of shock, amazement or heightened concern.

There are many uncertainties surrounding Arctic sea ice melt, among them the future pace of the melt, the year by which the Arctic may be ice-free in the summer, the balance of causes of the melt, and the impact on weather conditions in the northern hemisphere. The changes are happening much more quickly than anticipated just a few years ago, and the risks and opportunities are consequently more imminent.

The melt could have a wide range of negative impacts affecting the local ecology and local communities. Melting sea ice does not affect sea levels, but the warmer temperatures in the far north will speed the melting of Greenland's massive land ice sheet which would increase sea levels significantly. The carbon-rich permafrost beneath shallow coastal waters is more likely to melt releasing more GHGs into the atmosphere. Methane in sub-sea deposits is also more likely to be released. The relationship between Arctic sea ice melt and weather patterns is complicated, but some scientists believe there is enough evidence to show a link with atmospheric-pressure anomalies that make extended cold periods in Western Europe and North America more likely.

Methodology

We carried out content analysis of the three newspapers in each country by using the Lexis-Nexis and Factiva search engines for the UK, USA, and France, Newsbank for Australia, Atekst for Norway, and the papers' own websites' search facilities for India. For the first two IPCC reports, we looked at four days around their launch dates, and for the IPCC SREX report we examined four days both around 18 November 2011 (when there was some initial coverage of it) and its official launch date of 28 March 2012; and for the Arctic we looked at articles starting in January 2010 or (in the case of the UK in January 2011), and finishing on 30 September 2012.

We included Sunday editions of newspapers so for Australia the full list of papers examined is the *Australian/Weekend Australian, Herald Sun/ Sunday Herald Sun (Melbourne),* the *Sydney Morning Herald/Sun-Herald;* and for the UK, the *Guardian/Observer,* the *Daily Telegraph/Sunday Telegraph,* and the *Daily Mail/Mail on Sunday.*

In most cases, we did not include the online versions of the newspapers in our samples. The exception was India where we had to use search engines based on the newspapers' online sites. We are aware of the important limitations of not looking at online coverage in general and the echo chamber of social media linked to online stories in particular. It is well known that although there are important variations between countries, many, particularly younger, people in Western societies now consume information via websites and other 'new' media, and not print.[3] It is also the case that many more articles about climate change, and discussions about them, can be found on some newspaper websites than in the print version. This is particularly true of the *Guardian*, the *New York Times*, and the *Daily Mail* in our study.

However, despite declining circulations in many countries (but not India), print media are still read by vast audiences. The print editions of the 18 newspapers in this survey had a combined daily circulation in 2012 of more than 15 million, with a significantly higher readership.[4] Print editions still play a key role as agenda setters for what 'thought leaders' like politicians take seriously, for what is covered by journalists in other media and for what is discussed by the general public. And it is not unreasonable to assume that even though there is not a close fit in some cases between print and online coverage of the same topic, for most newspapers the print versions normally set the overall editorial tone and treatment.

The word searches gave us a total of 344 articles for the six countries, ranging from the lowest of 43 in Norway to the highest of 69 in the UK. The total number of articles by country, paper, and period is given in Table 5.2. In the case of the UK newspapers, for period 4 (the Arctic sea ice melt) we brought forward the starting date for the period under examination to 1 January 2011 (rather than January 2010) in order to maintain a rough parity between all six countries for the total number of articles

Table 5.2 Number of Articles by Country, Newspaper, and Period

	Period 1 IPCC WG-1	Period 2 IPCC WG-2	Period 3 IPCC SREX	Period 4 Arctic Sea	Total
Australia					
Australian	12	5	6	7	30
Herald Sun	4	3	2	3	12
Sydney Morning Herald	6	6	2	5	19
					61
France					
Le Monde	9	5	1	14	29
Le Figaro	8	3	1	7	19
Le Parisien	5	2	0	5	12
					60
India					
Business Standard	2	1	1	3	7
The Hindu	1	1	2	20	24
The Times of India	5	4	2	14	25
					56
Norway					
Aftenposten	6	0	3	16	25
VG	8	0	0	3	11
Dagbladet	2	3	0	2	7
					43
UK					
Guardian	7	2	1	22	32
Daily Mail	3	2	0	3	8
Telegraph	7	5	2	15	29
					69
USA					
New York Times	7	4	2	13	26
USA Today	5	3	1	5	14
Wall Street Journal	4	1	0	10	15
					55
Totals	101	50	26	167	344

reviewed. The *Guardian* in particular offers extensive coverage of the Arctic compared to the other newspapers in the sample, which would have inflated the UK sample significantly higher than the other five countries.

The texts of the two coding sheets for (1) the three IPCC reports and (2) the Arctic coverage can be found in Appendix 1. Detailed notes on the methodology we used can be found in Appendix 2. This includes details of how we measured different indicators of uncertainty, how quotes were counted, how sceptics were coded, and a full description of the IPCC concepts of likelihood and confidence levels. How we measured the relative presence, salience, and dominance of the four frames mentioned above (uncertainty, disaster/implicit risk, explicit risk, and opportunity) via the coding sheets can also be found in Appendix 2.

In several articles, headlines, opening paragraphs, and general content did not fit any of the four frames we had identified – in which case, they were registered as 'none' on the coding sheets. In a small number of cases, more than one frame was clearly competing strongly for dominance so we coded these articles as falling under two (or very occasionally three) frames. Given the focus of this study, we were particularly interested in the explicit risk frame, but it turned out that it was very seldom *the* dominant tone on its own. So we reviewed the articles where there was a significant presence of the explicit risk frame, and looked more closely at its presence as a dominant tone alongside one or more other frames. An illustrative example of (the unusual occurrence) of three frames together being dominant tones can be found in Appendix 2.

Our main aim was to test the extent to which the four different frames were present, salient or dominant across the 344 articles. In addition, we assessed (a) the number of articles where the explicit risk frame was combined with either the uncertainty frame or the implicit risk frame; (b) the extent to which the articles ignored the IPCC definitions of likelihoods either by not including them, or by including them but not defining them; and (c) the extent to which the articles included or emphasised more uncertainties about climate science than the mainstream scientific consensus would uphold by including sceptical voices.

Six country results

Table 5.3 summarises the results for the 344 articles from the six countries, and Figure 5.1 shows some of the same results in the form of a graphic.

Table 5.3 Headline Results of the Six Country Studies

Six country results	Period 1 WG-1	Period 2 WG-2	Period 3 SREX	Total IPCC	%	Period 4 Arctic	%	Total	%
Number of articles	101	50	26	177		167		344	
Uncertainty	**91**	**40**	**23**	**154**	**87**	**117**	**70**	**271**	**79**
More certainty	58	12	6	76	43	65*	39	n/a	n/a
Duelling experts	36	10	11	57	32	n/a	n/a	n/a	n/a
Salience	16	5	8	29	16	26	16	55	16
Direct quotes	55	23	19	97	55	64	38	161	47
Word presence	31	1	30	62	n/a	13	n/a	75	n/a
Sceptics	33	10	8	51	29	13	8	64	19
Dominant tone	21	8	14	43	25	39	23	82	24
Implicit risk	**83**	**48**	**24**	**155**	**93**	**126**	**75**	**281**	**82**
Salience	48	36	12	96	54	57	34	153	44
Direct quotes	62	40	22	124	70	71	43	195	57
Adjectives	63	30	11	104	59	55	33	159	46
Metaphors	19	6	1	26	15	30	18	56	16
Dominant tone	64	35	15	114	64	80	48	194	56

Explicit risk	**24**	**29**	**13**	**66**	**37**	**25**	**15**	**91**	**26**
Salience	0	7	3	10	6	8	5	18	5
Direct quotes	16	17	12	45	25	7	4	52	15
Adjectives	4	5	0	9	5	4	2	13	4
Metaphors	2	0	0	2	1	0	0	2	1
Word presence	48	73	27	148	n/a	46	n/a	194	n/a
Dominant tone	11	18	7	35	20	12	7	47	14
Opportunity	**10**	**10**	**0**	**20**	**11**	**72**	**43**	**92**	**27**
Salience	0	0	0	0	0	25	15	25	7
Direct quotes	6	6	0	12		14	8	26	8
Word presence	2	3	1	6		18	11	24	7
Dominant tone	1	2	0	3	2	38	23	41	12
IPCC concepts	**45**	**18**	**16**	**79**	**44**				
Explanation	22	1	4	27	15				

* Year ice-free

How the numbers in each row of Table 5.3 were arrived at from the coding sheets is laid out in Appendix 2. The individual country results can be found in Tables 6.1 to 6.6 in the following chapter. These are the six headline results:

- There was a consistent picture from the six countries of a very strong presence of the disaster/implicit risk and uncertainty frames compared to the explicit risk and opportunity frames. This was true for most of the climate change stories examined, and across the different media and political contexts of the six countries, and the range of newspapers.

- The disaster frame was present in 82 per cent of the 344 articles, making it the most common frame. For coverage of the three IPCC reports, it was present in over 90 per cent of them. It was also the most salient (in the headline or first few lines) with 44 per cent of the articles containing the frame, more than twice the next most common frame. It was also by some margin the most dominant tone of all four frames with well over half the articles containing it.

- Uncertainty was the second most common frame after the disaster frame. It was present in 271 of the 344 articles (79 per cent). However, judged by salience and dominant tone, we get a different picture. Only 16 per cent of the articles had uncertainty as salient, and only 24 per cent had uncertainty as the dominant tone.

- Opportunity was the third most common frame, being present in 92 or 27 per cent of the articles. However, these were overwhelmingly examples of the type of opportunities from not doing anything about reducing GHG emissions. Only five of the 344 articles (less than 2 per cent) in the total sample mentioned the opportunities from switching to a low-carbon economy.

- Explicit risk was the least present of the four frames, having a presence in 26 per cent of the 344 articles. It was also the least salient (5 per cent of articles). It was *the* dominant tone in just three articles. Its presence would have been more marked if the articles covering the three IPCC reports had included an explanation of the concepts of likelihood and confidence levels – only 27 of the 177 articles (15 per cent) did so. However, it had an above average presence, salience, and dominance in the coverage of the SREX report.

- Journalists follow the prompts from scientists and their reports: 70 per cent of the articles covering the IPCC reports, and nearly 60 per

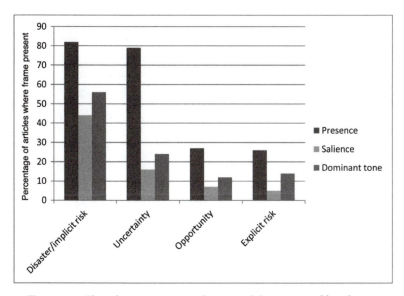

Figure 5.1 The relative presence, salience, and dominance of four frames

cent of all the articles in the sample, included quotes from scientists or scientific reports expressing some variant of the disaster/implicit risk frame. Nearly half of all the articles included a quote which indicated some manifestation of uncertainty. Nearly half the articles covering the SREX report contained explicit quotes about risk.

Discussion of the four frames

Disaster/implicit risk: a closer examination of the numbers in Table 5.3 shows how this frame was not only the most present, salient, and dominant, but also had the largest number of articles containing quotes from scientists or scientific reports expressing some variant of the disaster frame, the largest number of adjectives describing the adverse impacts and the largest number of articles containing metaphors capturing the frame. Table 5.4 shows that the ranges for presence and dominance between the six countries were not wide: presence came between a low of 74 per cent (Norway) and a high of 87 per cent (Australia); dominance between 49 per cent (Australia) and 64 per cent (India and the UK). The frame had the strongest presence in four of the six countries, easily the strongest

Table 5.4 Six Country Comparison for Four Frames

% of articles	Average for six countries	Australia	France	India	Norway	UK	USA
Uncertainty frame							
Presence	79	89	81	64	81	81	76
Salience	16	21	17	16	12	16	13
Dominant tone	24	25	32	27	16	16	27
Implicit risk frame							
Presence	82	87	81	82	74	84	80
Salience	44	56	58	30	37	46	36
Dominant tone	56	49	53	64	56	64	53
Explicit risk frame							
Presence	26	44	37	16	9	20	27
Salience	5	5	7	0	0	9	9
Dominant tone	14	11	19	13	5	13	20
Opportunity frame							
Presence	27	7	20	21	33	38	44
Salience	7	0	7	11	9	7	11
Dominant tone	12	2	10	14	16	9	24

salience in all six, and easily the strongest dominant tone in all six. In the two countries where it did not have the strongest presence (Australia and Norway), it was second to uncertainty by only a few percentage points (2 per cent and 7 per cent respectively).

It is not surprising that the disaster/implicit risk frame was so strong across all six countries in our sample. As mentioned in Chapter 4, other studies have shown that, in general, the frame is very common in media treatments of climate change over different periods. In our study, it was also in part a function of the selection of the three IPCC reports and the Arctic sea ice melt, which lent themselves to dramatic headlines and extensive treatment of the potentially adverse impacts of climate change. Journalistic norms of dramatisation and novelty clearly played a major role in the approach in each of the six countries we chose. But it is more significant that, as Table 5.3 shows, the disaster/implicit risk frame was more present, salient, and dominant in the reporting of the first IPCC report (WG-1) than in the second (WG-2), when the first report was about the science of climate change and the second report was about the impacts (and adaptation). As the British climate scientist Mike Hulme has written about the British newspaper treatment of the WG-1, the press 'offered embellished interpretations of what the consequences of future warming would be for the world. These impacts – invariably described as catastrophic, disastrous, and fearful – were to be the subject of the subsequent WG-2 report, but about these impacts WG-1 was completely silent' (Hulme, 2009b: 124).

Andy Revkin has written of the pull of the 'front page thought' which tempts journalists to overstate or play up the 'juiciest' facet of climate change stories, when the science is by nature 'incremental, contentious and laden with statistical analyses including broad 'error bars' (Revkin, 2005). His memory of reporting the WG-1 is that there was a journalistic pull to take the most dramatic quotes on the impacts from some of the IPCC lead scientists and headline them. It was a pull he resisted: 'It bugged me that some climate scientists went way beyond what the report said. The news was the report and not the spin.'[5]

Uncertainty: this frame featured strongly, as might have been expected given the vast array of complexities around climate science outlined in Chapter 2, and the way the coding sheets were set up to capture this. As explained above, the net was cast widely to include a wide range of uncertainty indicators from the use of words like 'could', 'may' and 'likely' to ranges of possible outcomes. However, it is worth pointing out that just under half of all the articles contained direct quotes from scientists or scientific reports expressing uncertainty. And in the first three periods, duelling experts were present in nearly a third of the articles.

Table 5.4 shows that there was not much variation between the six countries for this frame. In five of them, uncertainty was present in more than 75 per cent of the articles. India was the lowest with 64 per cent. It was the second most salient and second most dominant frame in each of the six countries. However, in each case the salience and dominance percentages were significantly lower than the presence percentages, in most cases by around 50 per cent.

In the case of the IPCC reports, the very high percentage of articles with an uncertainty presence (87 per cent) was tempered by the presence of the 'increasing certainty' frame in 76 (43 per cent) of them. And in the cases of the Arctic sea ice melt, much of the uncertainty in the reporting was focused on the year the Arctic might become ice-free in the summer months (nearly 40 per cent).

It is interesting to note that the word 'uncertain' was present 75 times across all the articles, but the actual usage was not a particularly useful pointer to the presence of uncertainty. For example, in the UK sample the word was only present in six of the articles, even though the concept was present in 56 of them. Also, half of the times it was used in all of the articles came from the Australia and US samples (37 out of 75); in Australia, 18 of the 20 times it appeared came from the reporting of the SREX report. It appeared in ten articles, whereas the concept was present in 54. In the USA, it appeared mostly in period 1, but only in a total of ten articles, whereas the concept was present in 42.

Sceptics were quoted or mentioned in 64 of the 344 articles, equivalent to 19 per cent. However, this figure masks some interesting country variations. Australia had the highest number of articles in the sample with sceptics in them (20) and the highest percentage of articles (33 per cent), followed by the USA in percentage terms (13 articles, 24 per cent). The UK came equal third with France (20 per cent) but had more articles. Norway was a long way behind (9 per cent), and India (2 per cent) even further. The fact that Australia, USA, and UK were the top three is broadly consistent with other studies. More detailed discussion can be found in the individual country sections. Forty-three of the 64 articles did give a clear indication of where the mainstream consensus could be found on climate science, and those that didn't were often opinion pieces by sceptics.

The IPCC's SREX report contained uncertainty, and both implicit and explicit risk within it, so it is interesting to mention how different newspapers reflected the balance between the frames. As the next chapter shows, in several articles in November 2011 the *Australian* focused on scientific uncertainty in its headlines, adjectives, and opening paragraphs, mentioning the word 'uncertain', 'uncertainty' or 'uncertainties' 16 times. In contrast, in its one article covering the report in November, the *Sydney Morning Herald* headlined its piece 'More wild weather on the way, UN climate panel says', although it too stressed the levels of uncertainty (Cubby, 2011). The *Guardian* also headlined the likely increase in more wild weather, and did not mention the uncertainties and difficulties of tying specific extreme weather events to global warming until the eleventh paragraph (Harvey, 2011). In contrast, the *Daily Telegraph* put the uncertainty (around the impacts) much higher up (Gray, 2011).

Opportunity: 72 of the 92 articles where this frame was present came from the Arctic sea ice period. This was to be expected given the obvious media interest in the opening of new shipping routes, and the possible mineral, gas, and oil exploration. All of the 25 articles where the frame was salient were type ii) opportunities (from doing nothing to reduce GHG emissions), as were 38 of the 41 articles where it was a dominant tone. It was present in 43 per cent of all the articles from the Arctic sea ice period. As you would expect, the frame was particularly present in the British (58 per cent), Norwegian (48 per cent) and US press (64 per cent) when compared to the Australian press (13 per cent), which was probably because of the geographical distance and the greater interest in the Antarctic. The *Wall Street Journal* had it as a dominant tone in five of its ten articles in our sample. In the other business-orientated newspaper journal in our sample, the *Business Standard* in India, it was also strongly present but in a small sample size (two out of three articles).

The frame was present in just 11 per cent of the articles covering the three IPCC reports. As already mentioned, most of these (15 out of 20) were manifestations of the opportunities offered by not doing anything about GHG emissions such as fewer colds and friendlier farming conditions in the northern hemisphere, or as the *Daily Mail* put it on 6 April 2007, 'a greener, lusher world'. Only five mentioned the possible advantages of new technologies, renewable energy or a low-carbon economy. The low presence of these opportunities could be due in part to our exclusion of the reporting of the third IPCC report of May 2007 (WG-3) on mitigation, which was much more orientated towards solutions. We did this in part because other studies showed it received much less coverage than the previous two reports (Hulme, 2009b: 121). However, it was of note that very few articles covering WG-1 and WG-2 mentioned the fact that a third report was coming along later which, amongst its conclusions, stated that a significant switch to cleaner technologies could cut expected temperature rises by half. This has prompted criticism of the IPCC that it might be advisable to highlight more prominently solutions and opportunities at the same time as stressing the potential negative impacts (ibid., 126).

Explicit risk: this was relatively often a dominant tone, found in 47 or 14 per cent of the articles but, in all but three examples, it was coupled with other frames, most usually with implicit risk. There were very few examples of a combination of uncertainty and explicit risk. Examples of the times when the explicit risk frame was combined with other frames, and what prompted the presence of the frame, can be found in the individual sections of the next chapter. In general, the presence of the frame was swollen by the use of the word 'risk', which occurred 194 times across all the articles – far more than the words 'opportunity' or 'uncertain'. For example, in Australia where the frame was the most present (in 44 per cent of articles), this was largely because of the frequent use of the word 'risk' (67 times or over a third of the appearances of the word in our sample). Indeed, on many occasions the simple use of the word was not a strong indicator of a generalised representation of explicit risk, as very often the concept was not developed.

The low presence of the explicit risk frame was in part due to the fact that the SREX report on extreme weather received relatively little coverage – only 26 articles of which ten were found in the Australian newspapers. One would have expected more coverage given the increase in extreme weather events in many countries, but several factors could have accounted for this (less deployment of IPCC resources, a leaked early version of the report, a general drop-off of interest in climate change stories, or the complexity of the main messages of the report). It was to be predicted that this period would contain a strong presence, salience, and dominance of the explicit risk frame given the presence of the word risk in the title of the SREX report, and the use of the word several times in the accompanying press release. Our results would partially bear this out: it was present in 13 of the 26 articles (50 per cent), compared to its overall presence for all four periods of 26 per cent, for period 1 of 24 per cent, and for the Arctic of 15 per cent (see Table 5.3). However, the frame was slightly more present in period 2 (29 articles out of 50, or 58 per cent). It was not particularly salient in period 3 (only three out of 26 articles, or 12 per cent), but it was relatively often present as a dominant tone (in 27 per cent of articles). Where its presence can be noted is in the number of articles with quotes from scientists or the actual report mentioning risk explicitly. Nearly half contained such quotes, which is significantly higher than the other periods.

Journalists are bound to be prompted by the presence of certain types of language or framing in press releases, summaries of reports or quotes, and this would seem to be the case with the SREX report. However, the presence of a frame can also be a function of the individual preferences of journalists writing the story. As we shall see in the next chapter, the *New York Times* was more likely than the other US newspapers to use the explicit risk framing. Of the nine articles covering the IPCC reports that included direct quotes from scientists using explicit risk language, seven were from the *NYT*; and of the 27 times the word 'risk' was used, 20 were also found in same newspaper.

It is interesting to note that, in general, the presence of the explicit risk frame would have been boosted if the articles *had* contained more explanation of what the IPCC concepts of likelihood and confidence levels meant in numerical terms. Forty-four per cent of the articles covering the three IPCC reports did include the concepts, but only 27 of the 177 articles (15 per cent) explained what they meant. The USA, Australia, France, and Norway were above the average figure, but India and the UK were well below. In these last two countries a total of only three of the 48 articles covering the three IPCC reports included the numbers.

A possible contributing factor to the generally low figure is that in the press releases which came out at the same time as the reports,[6] the numbers accompanying the likelihood and confidence levels do not seem to have been included, even though the concepts were used throughout the press releases. This was true of both the official IPCC press releases (for the WG-1 and the SREX reports),[7] and the press releases from the IPCC's parent organisations, UNEP and the WMO, who together established the IPCC in 1988.[8] However, they were explained in the Summary for Policy Makers, which many journalists were given and quoted (see for example Wynn and Doyle, 2007).

Finally, a brief mention should be made of two other interesting findings. The first is the relative presence of mentions of scientific versus societal impacts. A full explanation of the difference is given in Appendix 2, but examples of scientific impacts would be species loss, rainforest 'dieback', and the changing ecology of the Arctic, whereas societal impact would be mass migration, droughts or floods affecting people, or food shortages. It is worth pointing out that in the reporting of the three IPCC reports, there was far more emphasis placed on societal impacts than scientific impacts (116 articles versus 30, the rest being unclear). This would be expected of the WG-2 with its focus on impacts, but it was also true of WG-1 which was focused on the science (52 articles versus 23). This may have been another example of journalists pushing beyond the science headlines of the report to make it more relevant to their readers. Most of the impacts found in the coverage of the Arctic sea ice were scientific. But it is interesting that there were considerably more articles with mentions of the impact on the local ecology (mostly polar bears) than the impacts on the local Inuit people (47 versus 16).

Secondly, there has been considerable debate over whether it is a positive approach for scientists to express emotion when interviewed by the media, in that this can in some circumstances engender more trust with the viewer or reader. For example, the Shuckburgh study suggests that focus groups were happier to accept a statement of uncertainty if it was combined with an emotional statement (such as fear) (Shuckburgh, Robison and Pidgeon, 2012: 39). The counter-view is that people expect scientists to be dispassionate so that emotions do not get in the way of the science. For the record, of the 167 articles about the Arctic summer sea ice melt, a total of 26 had a scientist or scientists expressing an emotion about the disappearing summer ice. Of the first such quotes in these articles, 13 expressed surprise, seven worry, four shock, and two fear.

6

Country Studies

Australia
by Lyn McGaurr and Libby Lester

The story of Australia's media coverage of climate change from 2009 onward is intimately associated with protracted and intense political controversy about how the country should deal with its carbon emissions and its international obligations.

When Australia signed the Kyoto Protocol in 1998, it was one of the highest per capita emitters of GHGs in the industrialised world and one of the few industrialised signatories permitted to increase its emissions under the agreement. At last report, the country was on track to meet its Kyoto obligation of limiting emissions to an average of 108 per cent of 1990 levels over the period 2008–12 (Department of Climate Change and Energy Efficiency, 2012: 15). An economy that owed much of its strength through the global financial crisis to its mining industry, in 2012 it had the highest per capita GHG emissions in the developed world and was the 15th largest emitter overall (Flannery, Beale and Hueston, 2012: 4).

According to a 2007 poll by the independent think tank the Lowy Institute, Australians rated climate change the most important external threat the country faced (Gyngell, 2007: 2). In response to this groundswell of popular concern, the Coalition Government and the Labor Opposition of the day promised an emissions trading scheme as part of their 2007 election campaign platforms, and in December that year Labor's Kevin Rudd ratified the Kyoto Protocol in his first official act as the country's new prime minister (*Sydney Morning Herald*, 2007).

In August 2009, legislation for an emissions trading scheme was rejected by Parliament, but Rudd remained committed to action, reiterating to the General Assembly of the United Nations in September his earlier claims that climate change was the 'great moral challenge of our generation' (Rudd, 2007). The government revised the legislation, in negotiation with Opposition leader Malcolm Turnbull. However, before its passage could be secured, Turnbull lost a leadership ballot by one vote and the Opposition withdrew its support for the government's scheme, contributing to the legislation's defeat in early December 2009. Later that month Rudd fought hard for international agreement at the Copenhagen summit but was disappointed with the outcome. Soon after, he postponed further attempts to introduce an emissions trading scheme of his own. Julia Gillard, who became Labor Party leader and prime minister in June 2010, went to an election two months later promising not to introduce a carbon tax (*Sydney Morning Herald*, 2010) but, when she found herself requiring the support of the Greens and Independents to form a government, she agreed to put a price on carbon as an interim measure. Legislation for a carbon tax was passed in November 2011, and the tax was introduced in July 2012. The Opposition has said it will remove the tax if it wins government at the next election due in September 2013.

Today Australia is committed to reducing its GHG emissions to 5 per cent less than 2000 levels by 2020 and has said it will consider more ambitious targets for that period if the global community takes similar action (Flannery, Beale and Hueston 2012: 8). However, opinion polls suggest that the public's willingness to endorse strong action declined during the many years of political argument about how reductions should be achieved. By 2012 only 36 per cent of Australians supported the statement that global warming was a serious problem requiring action even if that action came at significant cost (Hanson, 2012: 6). At the same time, 45 per cent agreed with the statement 'The problem of global warming should be addressed, but its effects will be gradual, so we can deal with the problem gradually by taking steps that are low in cost', up from 24 per cent in 2006 (ibid.). Decreases in concern were also evident in Gallup polls. These identified a decline between 2007–8 and 2010 in Australians who viewed climate change as a very or somewhat serious threat from 75 per cent to 68 per cent (Pugliese and Ray, 2011).

The apparent decline between 2006 and 2012 in Australians' willingness to take strong action to mitigate global warming came about in a period of sustained media interest in climate change policy. In 2009 for example, climate change was the single biggest issue when all media were counted, and the second biggest press issue after the global recession (Baume, 2009). In 2012 the carbon tax ranked third when all media were counted, and fourth in the press (Chalmers, 2012). In a study of articles in ten Australian newspapers from February to July 2011 conducted by the Australian Centre for Independent Journalism (ACIJ), News Ltd's broadsheet the *Australian* – the country's only national generalist newspaper – was found to have published more climate change articles than other newspapers in the study: 28 per cent of total articles counted appeared in the *Australian*, compared, for example, with 13 per cent in the *Sydney Morning Herald* and 9 per cent in the *Herald Sun* (Bacon, 2011).

A study of news articles published by Fairfax Media's *Sydney Morning Herald* and News Ltd's *Courier Mail* in July 2007 found climate change was generally represented as caused by humans (Howard-Williams, 2009: 37). In addition, since the early 2000s the *Australian* has described itself as accepting the core science of global warming (Manne, 2011). However, these details belie a more complex media representation, as borne out by the current study. In many of the news articles in the *Sydney Morning Herald* and *Courier Mail* in July 2007, for example, climate change was presented as being 'the result of unspecified emissions [...] implying a human contribution to climate change while not providing any specific link to how or why these emissions came about' (Howard-Williams, 2009: 36). Many articles 'advocated solution through progress and technological advancement supported by government and industry' (ibid., 37). However, in the *Australian*, 47 per cent of articles about climate change policy in the ACIJ study noted earlier were found to be negative, 44 per cent neutral, and 9 per cent positive (Bacon, 2011: 12) – although when the report was made public the *Australian* responded by challenging its lack of criteria for identifying stance (Jones, 2012).

Furthermore, in spite of the *Australian*'s editorial policy of acknowledging the human causes of climate change, a separate analysis by political scientist Robert Manne of the climate change articles it published between January 2004 and April 2011 found that its editorials often denigrated those in favour of radical action (Manne, 2011: 47) and its news items and opinions page were 'overbalanced by a factor of four to one by writers unfavourable to action' (ibid., 54). This study also found that the *Australian* published 'scores of articles by people who claimed to know that the consensual view of the climate scientists was entirely bogus but who have not passed even a first year university examination in one of the relevant disciplines' (ibid., 52). Sceptics were given a particularly prominent place in its opinion pages. In addition to columns by sceptical non-scientists, this report found that the *Australian* published dozens of opinion pieces by scientists challenging the consensual view of climate change, including Ian Plimer, Bob Carter, Lord Monckton, and David Bellamy. The study found that these pieces outnumbered opinion columns by climate scientists supporting the consensus by ten to one.

Although the *Australian's* circulation is relatively small – 122,000 Monday to Fridays and 267,000 on Saturdays in late 2012 – it is regarded as having significant influence both within the political class (McGaurr and Lester, 2009: 180) and among other newspapers in News Ltd's Australian stable (Manne, 2011: 4). One of these, the Melbourne-based *Herald Sun*, is the country's top-selling daily newspaper, with a circulation of 450,000 Monday to Fridays and 444,000 on Saturdays. At the time the current study was conducted, this tabloid did not appear to have an environment editor but videos and, occasionally, articles by the *Australian's* environment editor, Graham Lloyd, could be accessed via its website (www.heraldsun. com.au). A regular columnist, Andrew Bolt, is a vocal climate science sceptic who also appears in Adelaide's *Advertiser*, Sydney's *Daily Telegraph*, *The Bolt Report* on the Ten Network, and in his own blog, which is promoted as 'the most-read political blog in Australia' (Bolt, 2013). In the 2011 ACIJ study, columnists Bolt and another sceptic, Terry McCrann, were found to have published the most individual opinion pieces about climate change policy in the ten papers reviewed (Bacon, 2011: 57–8).

Like the *Australian*, the *Sydney Morning Herald* is aimed at an educated, generally affluent demographic, but its editorial stance has traditionally been more liberal. Published by Fairfax Media, in late 2012 it had a circulation of 161,000 Monday to Fridays and 265,000 on Saturdays. In the ACIJ study, the *Sydney Morning Herald* and its sister newspaper Melbourne's *Age* had a more even distribution between positive and negative opinion pieces than those in News Ltd papers (ibid., 16, 55).

Despite Australia's relatively small population it has a strong international reputation in climate science, evident in the fact that it has contributed four co-ordinating lead authors and eight lead authors to Working Group 1 of the IPCC's Fifth Assessment Report (n.d.). In recent years, some Australian climate scientists have become concerned about the way climate change is being reported in the media. In August 2009 for example, 15 of the country's most eminent climate scientists put their name to an opinion piece stating their views about the seriousness of the risks posed by climate change (Raupach et al, 2009). In a journalism forum broadcast by the ABC's Radio National in November 2010, one of these scientists, Professor Anne Henderson-Sellers of Macquarie University, expressed extreme frustration at the publicity given to the views of climate sceptics by the Australian media. In response to a question about uncertainty she also spoke in terms of the IPCC's levels of likelihood, arguing that it would be better for the media to talk of levels of confidence than uncertainty (Henderson-Sellers, 2010).

In 2011 the Australian government established the independent Climate Commission, headed by Professor Tim Flannery, a scientist with a high public profile as a climate science communicator, 'to provide Australians with an independent and reliable source of information about climate change' (Climate Commission, 2012: 1). During its short history to date, the Commission has released a number of publications seeking to help the general public and the media better understand the risks and uncertainties of climate change. Its report, *The Critical Decade: Climate Science, Risks and Responses* (2011), was released two and a half years after Australia's worst bushfires had killed 173 people, and just a few months after devastating floods had affected 200,000 and killed 35. The report stated that climate change was becoming increasingly contested and being 'attacked in the media by many with no credentials in the field' even as the scientific understanding was continuing to advance (ibid., 3). Although in its 60 pages it mentioned 'risk' or 'risks' 57 times, it also mentioned 'uncertain', 'uncertainty' or 'uncertainties' 43 times. In January 2013, however, during Australia's hottest summer on record (Bureau of Meteorology, 2013) and following its most severe heatwave since monitoring began, the Climate Commission released a five-page brochure for 'citizens and media seeking to understand the link between climate change and the very unusual weather' (Karoly, England and Steffen, 2013: 2). Here 'risks' were mentioned five times but 'uncertainties' not at all.

The Australian media's coverage of risk and uncertainty

The results of our coding can be seen in Table 6.1. The results for the coverage of the three IPCC reports would suggest the following:

- The **uncertainty** frame was present in 42 of the 46 articles, but it was only salient in a large proportion of the articles in period 3, when it appeared in a headline or was dominant in the first few lines in seven of the ten articles. Interestingly, period 3 was also the only one in which the uncertainty frame was the dominant tone, registering as

Table 6.1 Australia Results

	Period 1 WG-1	Period 2 WG-2	Period 3 SREX	Total IPCC	%	Period 4 Arctic	%	Total	%
Number of articles	22	14	10	46		15		61	
Uncertainty	**20**	**12**	**10**	**42**	**91**	**12**	**80**	**54**	**89**
More certainty	10	1	1	12	26	6*	40	n/a	n/a
Duelling experts	14	5	6	25	54	n/a	n/a	n/a	n/a
Salience	3	0	7	10	21	3	20	13	21
Direct quotes	16	11	10	37	80	8	53	45	74
Word presence	0	1	18	19	n/a	1	n/a	20	n/a
Sceptics	11	3	3	17	37	3	20	20	33
Dominant tone	2	2	8	12	26	3	20	15	25
Implicit risk	**20**	**14**	**10**	**44**	**96**	**9**	**60**	**53**	**87**
Salience	9	14	3	26	57	8	53	34	56
Direct quotes	14	12	10	36	78	6	67	42	69
Adjectives	12	10	3	25	54	4	27	29	48
Metaphors	7	4	1	12	26	7	47	19	31
Dominant tone	9	10	3	22	48	8	53	30	49
Explicit risk	**9**	**8**	**6**	**23**	**50**	**4**	**27**	**27**	**44**
Salience	0	2	0	2	4	1	7	3	5
Direct quotes	3	4	4	11	24	0	0	11	18
Adjectives	0	0	0	0	0	0	0	0	0
Metaphors	0	0	0	0	0	0	0	0	0
Word presence	23	26	12	61	n/a	6	n/a	67	n/a
Dominant tone	1	3	1	5	10	2	13	7	11
Opportunity	**1**	**1**	**0**	**2**	**4**	**2**	**13**	**4**	**7**
Salience	0	0	0	0	0	0	0	0	0
Direct quotes	0	0	0	0	0	1	6	1	2
Word presence	1	3	1	5	n/a	0	n/a	5	n/a
Dominant tone	0	1	0	1	2	0	0	1	2
IPCC concepts	**10**	**6**	**8**	**24**	**52**	**n/a**			
Explanation	5	1	2	8	17	n/a			

* Year ice-free

such in eight of the ten articles (including one instance in which it shared dominance with implicit and explicit risk framing).

In period 1 it was the dominant tone in just two of 22 articles and in period 2 in only two of 14. In period 1, half of the 20 articles with an uncertainty frame were tempered by the 'increasing certainty' frame, but in the other two periods the figures were small – just one in 12 and one in ten respectively. The figure of 26 per cent of all articles in the Australian IPCC samples with an uncertainty frame being tempered by an 'increasing certainty' frame is low compared to the results in other countries.

Sceptical comment was present in half the articles in period 1, 21 per cent in period 2 and 30 per cent in period 3, while duelling experts were evident in more than half the articles in periods 1 and 3 and a little less than half those in period 2. The word 'uncertain' appeared 18 times in the ten articles in period 3 but did not appear at all in period 1 and appeared only once in period 2.

- The **implicit risk** or disaster frame was particularly present, as it was found in 44 of the 46 articles (96 per cent), but was only relatively salient (57 per cent of the articles). Interestingly, there were stark differences in salience between period 2, where the implicit risk frame was salient in all 14 articles, and periods 1 and 3, where it was salient in only nine of 22 articles and three of ten articles respectively. This was largely due to the *Australian*'s reporting. There was a strong reliance in periods 1 and 2 on the language of disaster, and the use of adjectives such as 'dire' and 'devastating', particularly in period 2. It was dominant in 48 per cent of the articles.

- The **explicit risk** frame was present in 50 per cent of the articles, mainly because of the high use of the word 'risk', which appeared 61 times in the 46 articles. Explicit risk was salient in only two articles, and a dominant tone in only five, in four of which it was equally dominant with other frames.

- The **opportunity** frame was almost absent.

- The **IPCC concepts** of likelihood and probability were present in 24 of the 46 articles, but were explained in only eight.

Results from the coverage of the Arctic sea ice melt are as follows:

- The **uncertainty** frame was present in 12 of the 15 articles (80 per cent), and in six of these this related in part or in total to questions about when the Arctic sea would be ice-free. It was salient in three articles, and dominant in the same number. However, more than half the articles for the period (eight of 15) contained uncertainty quotes.

- The **implicit risk** frame was present in nine articles (60 per cent), and was salient in eight. It was also the dominant tone in eight articles. However, although implicit risk was evident in adjectives and metaphors, these tended to be more subdued than those typically associated with disaster framing. Moreover, some were suggestive of both risk and uncertainty – for example, 'the implications are enormous and mysterious', 'uncharted territory' and the polar bear is 'the canary in the coal mine'.
- The **explicit risk** frame was present in four articles, and again this was significantly influenced by the high presence of the word 'risk'. Although 'risk' appeared six times, four of those appearances were in a single article, and in this article explicit risk was a dominant tone in combination with implicit risk. Explicit risk was also dominant in combination with implicit risk in one other article.
- The **opportunity** frame was present in only two articles, possibly due to a perceived lack of interest among Australian audiences in the economic fortunes of Arctic shipping and mining.
- There was little interest in the specific possible impacts. Most interest was given to possible changes in temperature (three articles).
- In two of the 15 articles scientists expressed surprise.

The results for all four periods

Reviewing all the articles, it is clear that there is a **very strong presence of uncertainty** overall (89 per cent of articles). A significant amount of the uncertainty is due to the inclusion of ranges of projections as indicators, but duelling experts and sceptical voices were also strongly present, particularly in the coverage of the first IPCC report (period 1). However, these factors alone do not account for the overwhelming predominance of uncertainty in the coverage of the SREX report (period 3), where it was the dominant tone in eight of the ten articles reviewed. Notably, uncertainty was the dominant tone in all six articles published in the *Australian* that were included in this period (including one in which uncertainty shared dominance with implicit and explicit risk).

A range of factors came together in the period around the SREX report (November 2011 to March 2012) in Australia:

- The Victorian fires of 2009 and the Queensland floods of 2011 had been devastating, making extreme weather a subject of public interest.
- As mentioned above, in mid-2011, some months after the floods, the Climate Commission had released a report mentioning 'risk' or 'risks' 57 times and 'uncertain', 'uncertainty', or 'uncertainties' 43 times. That is, uncertainty as well as risk was prominent, despite the document's stated objective of responding to 'the noisy, confusing "debate" in the media' by contextualising uncertainty to show that 'there is much science that is now confidently understood and for which there is strong and clear evidence' (Climate Commission, 2011: 3).
- A few days before the leak of the IPCC's SREX report a carbon tax had been passed into law following a protracted period of political debate in the media.
- In the IPCC's SREX report, uncertainty as well as risk was prominent.

In period 1 the *Australian* did not favour the implicit risk frame as much as the other two newspapers – in fact, a wide variety of tones were dominant in its articles. In both period 1 and 2, the paper published editorials arguing for a cool-headed approach. For example, on 9 April 2007, it wrote: 'The latest IPCC report has heightened community awareness and concern over climate change. Governments have a responsibility to act, but they must favour solutions that deliver real benefits over unrealistic responses that offer little, but grab headlines and exploit community fears'.[1] The *Australian*'s position by period 3 was articulated by its editor, Chris Mitchell, on 2 December 2011 in an article rebutting the ACIJ study cited earlier: 'The carbon tax is one of the biggest concerns for business in this country and it is only proper that a newspaper such as ours reports those concerns' (Leys, 2011).

When the IPCC's SREX report was first reported in November, the *Australian* focused on scientific uncertainty in headlines, adjectives, and opening paragraphs. It was also responsible for 16 of the 18 mentions of the words 'uncertain', 'uncertainty' or 'uncertainties' in the period 3 sample. Examples of headlines at the time were 'Climate change effects unknown', 'Review fails to support climate change link', and 'Abbott jumps on great unknown'. In contrast, these words were not mentioned in any of the reviewed articles from any of the newspapers in period 1, and there was only a single mention, in the *Sydney Morning Herald*, in period 2.

As expected, the **implicit risk** frame was present in nearly all articles reviewed as part of the current study (87 per cent of articles). It was dominant in nearly half the articles and was especially salient in period 2 (the second IPCC report), where it was salient in all 14 articles. It was also highly salient in period 4 (the Arctic sea ice melt). Adjectives that suggested disaster were common in the articles in period 2.

Explicit risk, on the other hand, had low salience and dominance. It was a dominant tone on only seven occasions, and only on one of those was it the only dominant frame. On the other occasions it shared a dominant tone with other frames, and particularly implicit risk. However, it was often strongly present through use of the word 'risk'. In period 1 mentions of this word were mostly concentrated in two articles, one of which had a dominant frame of uncertainty, despite the presence of the 'increasing certainty' frame and a stated acceptance that action was warranted. This article, which appeared in the *Australian*, used the word 'risk' six times, mentioned the idea of insurance, and used and explained the IPCC's concepts of likelihood (Warren, 2007: 19, 29). In period 2, mentions of risk were largely concentrated in four of the 14 articles, while in period 4 a single article contained four of the six mentions. This concentration suggests that the use of the word is not necessarily a strong indicator of a generalised representation of explicit risk across the media.

Interestingly, our results show that explicit risk would have been more strongly present if newspapers *had* explained their use of the IPCC's levels of likelihood and confidence more often. The terms were used extensively in coverage of the IPCC's reports – particularly in period 3, where they were present in eight of the ten articles – but they were very rarely explained except in period 1, where they were explained on five of the ten occasions they were used.

As expected, the **opportunity** frame was almost entirely absent in the current study, even being rare in the coverage of the Arctic sea ice, when it was reasonably prominent in other countries. However, since the Climate Commission's establishment in 2011, it has used the language of opportunity in a number of its regional reports, three of which have the word in their title. The complete absence of the word 'uncertainty' from another report by the Commission – this one released in early March 2013 and entitled *The Angry Summer* (Steffen, 2013) – raises the possibility that in some quarters there has been a move away from writing in terms of uncertainty towards a stronger focus on levels of confidence.

France

by Toussaint Nothias and James Painter

In many significant and interesting ways, France is dissimilar to the other industrialised countries included in this study. Climate change is not subject to the same degree of intense political controversy as it is in Australia, the USA, and the UK; climate sceptics are present in the media and in the wider political setting, but are nothing like as consistently vocal a force; opinion polls normally show that the French public is more convinced by the science of climate change and the need to do something about it than in the UK and the USA; the media landscape in France is different to that of the UK – circulation of national newspapers, for example, is significantly lower, and there is no tabloid press similar to that in the UK; and climate change coverage is far less studied by academics than their counterparts in Australia, the UK, and the USA.

Unlike in the United States, but as in the UK, the belief in the importance of climate change, and the need to tackle it, is not so much of a defining right–left issue for French politics. The centre-right two-term president from 1995 to 2007, Jacques Chirac, was a strong advocate of tackling global warming within France, the EU, the G-8 and the UN. For example, in July 2006, at the G-8 meeting in St Petersburg in Russia, Chirac famously warned that 'humanity is dancing on a volcano' and called for more serious action by other G-8 members.

In similar fashion, the centre-right president who replaced him, Nicolas Sarkozy (2007 to 2012), also pushed hard for international agreements, and was very critical of those countries which stood in the way of a more ambitious deal at the Copenhagen summit in December 2009. During the summit, France had launched a joint initiative with Ethiopia (representing Africa) to adopt an ambitious agreement limiting the increase of temperatures to 2°C above pre-industrial levels, and ensuring that vulnerable countries would receive adequate financing to face the challenge. But in the rapid denouement of the summit, France and the EU were largely sidelined from the last-minute negotiations of the Copenhagen Accord, which revolved around the USA, China, Brazil, South Africa, and India. However, Sarkozy continued to push for international action on climate change, including a proposal, promoted in March 2010, to introduce a small tax on financial transactions in order to raise billions of dollars to help developing nations fight climate change.

Domestically, France has broadly followed the EU policy of aiming to reduce greenhouse gas emissions to 20 per cent below 1990 levels by 2020 and, in the longer term, reducing them by 75 per cent by 2050. Figures suggest that it managed to cut emissions by 3 per cent between 1990 and 2010, well below the EU average of 15 per cent (Byles, 2012, citing House of Commons Library statistics). In September 2009, Sarkozy launched a carbon tax plan, which proposed a charge of €17 (US$22) on each tonne of CO2 emitted by businesses and households, which could have generated US$4 billion a year. The tax only applied to oil, gas, and coal consumption, not to electricity, in part because around 80 per cent of France's electricity is generated by nuclear power, which does not emit GHGs.

The proposals met with opposition from green groups, who said the plans were not ambitious enough, and by parts of his party, the UMP, who said the plans went too far. The plan was also rejected by a constitutional court in December 2009, which ruled that there were too many exemptions for polluters in the tax plan. The court, which acted as a legal compliance watchdog, said 93 per cent of industrial emissions, other than fuel use, would be exempt from the tax. Sarkozy was never able to muster enough political support to pass the carbon tax, in part because opinion polls showed growing opposition in the face of other economic priorities such as tackling low growth and growing unemployment. Unlike in Australia, climate change was not explicitly a major issue in the 2012 elections, even though nuclear power and the future of energy did figure quite strongly. The new centre-left president, François Hollande, stressed the importance of a new international deal to come into force when the current Kyoto Protocol expired at the end of 2015, and warned of an environmental 'catastrophe' if countries did not invest in renewable energy. However, climate change did not appear to be a policy priority.

Despite public opposition to Sarkozy's carbon tax, French public opinion has tended to show less scepticism about climate change than the UK and – in principle – more support for action to tackle it. The widely respected Eurobarometer surveys of climate change opinions across the EU member states show that in 2009 81 per cent of the French people consulted thought that climate change was a very serious problem, and 51 per cent the *most serious* problem facing the world, in both cases above the average for the EU.[2] This contrasts quite sharply with the UK, where 'only' 51 per cent thought it was a very serious problem, the second lowest in the EU (after Estonia). A follow-up survey in 2011 showed that in France the level of concern remained pretty much constant.

One indicator of scepticism is to measure the degree to which people believe the seriousness of climate change has been exaggerated. In the 2009 survey, 31 per cent of the French public consulted agreed that it had been, compared to 40 per cent in the UK. Another way of measuring it is to ask people whether they believe that climate change has been proven by science. According to a 2012 survey of 13,000 internet users in 13 countries, France showed a slightly higher belief in this statement – 69 per cent compared to 63 per cent in the UK (but lower than Germany where 82 per cent believed it).[3]

Part of the reason that French people express less doubt about climate change than in the UK or the USA may be linked to the fact that sceptical views are only found on the fringes of the main political parties. As has been shown by several studies, a main driver of opinion on climate science is 'elite cues', or the presence of sceptical voices within the main political parties who are often quoted in the media (Brulle et al, 2012). A section of the US Republican Party often linked to the Tea Party has many such voices, while the Conservative Party in the UK was becoming increasingly riven by splits on the issue in 2011–12. In contrast, the official position of the right-wing National Front, which is still a strong force in French politics, does not question the basics of climate science, and is in favour of reductions in greenhouse gases, which marks it out from the British National Party and UKIP in the UK. There are some sceptic voices in the party leadership but they do not tend to appear in the media, unlike Lord Monckton in the UK (who is leader of the Scottish UKIP). The French nuclear industry has strong links with both centre-right parties and the Socialist Party, and has been a strategic industrial choice by the whole political class. This leaves little political space for anything comparable to some of the lobby groups in the UK, USA, and Australia which have links to fossil fuel or mineral extraction industries.

The 2011 RISJ study on climate scepticism showed that the French print media examined in the survey (*Le Monde* and *Le Figaro*) in general included far fewer sceptical voices than their counterparts in the USA and UK (Painter, 2011: chapter 4). For example, in the period November 2009 to February 2010, whereas only 5 per cent of the articles about climate change in the French newspapers included sceptics, the equivalent figure for the two UK newspapers was nearly 20 per cent and for the two US newspapers nearly 35 per cent. *Le Figaro* for example did include opinion pieces written by the prominent climate sceptic and former minister under the left-wing Prime Minister Lionel Jospin, Claude Allègre. But in general, sceptical voices did not get much play.

One of the reasons for this is that in France there is no equivalent of the UK tabloid press, where much of the most strident scepticism is found. A significant proportion of this is published in the opinion columns of the right-leaning *Daily Mail*, the *Sun*, and the *Daily Express*, which together have a circulation of more than 5 million. In contrast, the left-leaning *Le Monde* and the right-leaning *Le Figaro* have more of an 'elitist audience', with a circulation of around 330,000 for each of them in 2011.[4] The third paper included in this study, *Le Parisien*, is the sister publication of *Aujourd'hui en France*. Both are more like the mid-market tabloid in the UK, the *Daily Mail*, and appeal to a similar kind of audience. In 2011, the combined circulation was around 460,000, which would make them the best-selling national title in France if they were a single title. However, the regional newspaper, *Ouest-France* has the largest circulation of any French paper, of about 780,000.

Le Monde and *Le Figaro*, together with the third elite newspaper, *Libération*, dominate the elite market. Some argue this domination by just three well-established newspapers can mean that together they set a *de facto* agenda on coverage of an issue, including that of how much space to give to sceptical viewpoints. *Le Monde*'s science journalist, Stéphane Foucart, says that he only tends to quote articles that have been published in peer-reviewed science journals, which would help to explain the near absence of sceptics in his paper (cited in Painter, 2011: chapter 5.3). For her part, *Le Figaro*'s science journalist Marielle Court says that it was difficult to report on scientific uncertainties during the period of Climategate, which began in November 2009 when scientists at the climatic research unit at the University of East Anglia in the UK were accused of manipulating results and keeping critics out of science publications. She says the difficulties arose because of the powerful TV presence of such sceptics as Allègre, but these difficulties have abated.[5]

The output of these newspapers has not been subject to the same level of academic scrutiny that they would have had if they were being published in the UK, the USA or Australia. Of the studies that have been written, a survey of 144 articles in *Le Monde* and *Le Figaro* from 2001–7 concluded that the coverage of climate change came mostly in the form of longer news pieces that 'offered due background information' and promoted scientific certainty about climate science (Boyce and Lewis, 2009: 208–9). Another study published in 2004 looked at the difference between US and French media coverage between 1987 and 1997 and found that, whereas the former focused on conflict between scientists and politicians, the latter focused not on the scientific controversy but more on the dispute between the EU and the USA over international (in)action on climate change (Brossard, Shanahan and McComas, 2004). The French coverage was also more event-based, and presented a more restricted range of viewpoints on global warming than the American coverage. This showed the importance of the different domestic political and journalistic environments in which the media in the two countries operate. In France, journalistic culture such as a greater tendency to opinion rather than objectivity in the reporting was 'clearly reflected in the coverage and may have influenced the themes covered and the sources cited'.

Finally, a study published in 2012 looked more at the drivers of media coverage of climate change over the last 20 years to explain why there was more attention paid to the issue in the 2000s (Aykut, 2012). They argue that although a major storm in the winter of 1999 and the infamous heatwave in France and Europe in 2003 were important reasons for increased coverage, the main drivers were social shifts in four social or political sectors – journalism, science, the state, and NGOs. So in the journalistic field, environmental journalism became more of a fully-fledged speciality within newsrooms, which led to less coverage of political controversy. In the world of climate science, the scientists became more 'professionalised' in dealing with the press, while, as in many Western countries, NGOs stepped up their activities aimed at increasing the salience of the climate change issue.

The French media's coverage of risk and uncertainty

The five key hypotheses outlined in Chapter 5 were tested in our examination of the three French newspapers. The results summarised in Table 6.2 would suggest the following for the coverage of the three IPCC reports:

- The **uncertainty** frame was present in 29 of 34 articles (85 per cent). However, this was balanced by the 'increasing certainty' frame found in a significant number (21) of those articles. It was not particularly salient (only six of the articles had it as a headline or dominant in the opening few lines), but it was found as a dominant tone in a quarter of the articles (nine articles). It was also to be found in several direct quotes (in 13 of the articles), and in the presence of duelling experts (seven of the articles). It is worth noting that around a third of the articles had some mention of sceptical opinion (ten of the articles).

Table 6.2 France Results

	Period 1 WG-1	Period 2 WG-2	Period 3 SREX	Total IPCC	%	Period 4 Arctic	%	Total	%
Number of articles	22	10	2	34		25		59	
Uncertainty	20	7	2	29	85	19	76	48	81
More certainty	16	4	1	21	62	12*	n/a	n/a	n/a
Duelling experts	6	0	1	7	21	n/a	n/a	n/a	n/a
Salience	5	1	0	6	18	4	16	10	17
Direct quotes	11	1	1	13	38	11	44	24	41
Word presence	5	0	2	7	n/a	2	n/a	9	n/a
Sceptics	6	3	1	10	29	2	8	12	20
Dominant tone	5	2	2	9	26	10	40	19	32
Implicit risk	16	10	2	28	82	20	80	48	81
Salience	12	8	2	22	65	12	48	34	58
Direct quotes	13	6	2	21	62	10	40	31	53
Adjectives	13	5	0	18	53	11	44	29	49
Metaphors	6	1	0	7	21	6	24	13	22
Dominant tone	15	5	1	21	62	10	40	31	53
Explicit risk	7	5	2	14	41	8	32	22	37
Salience	0	1	1	2	6	2	8	4	7
Direct quotes	3	4	1	8	24	2	8	10	17
Adjectives	1	2	0	3	9	0	0	3	5
Metaphors	0	0	0	0	0	0	0	0	0
Word presence	8	5	2	15	n/a	8	n/a	23	n/a
Dominant tone	2	2	1	5	15	6	24	11	19
Opportunity	1	1	0	2	6	10	40	12	20
Salience	0	0	0	0	0	4	16	4	7
Direct quotes	1	0	0	1	3	4	16	5	8
Word presence	0	0	0	0	n/a	1	4	1	n/a
Dominant tone	0	0	0	0	0	6	24	6	10
IPCC concepts	8	2	1	11	32	n/a		n/a	
Explanation	5	0	1	6	18	n/a		n/a	

* Year ice-free

- The **implicit risk** or disaster frame was also particularly present and found in 28 of the 34 articles (82 per cent). It was particularly salient with nearly two thirds of the articles having it as a headline or dominant in the opening lines (22 articles). It was a dominant tone in 21 articles – more than twice as many as the uncertainty frame. In addition, it was the most quoted frame with quotes found in 21 articles (62 per cent). Strong adjectives reinforcing this 'implicit risk' frame – such as 'extrêmement grave', 'terrible', 'catastrophique', 'sombre' – were quite common and found in more than half of the articles (53 per cent).
- For its part, the **explicit risk** frame was significantly less present (fewer than half of the articles). Although it can be found in 14 articles (41 per cent), this was mainly related to the use of the word 'risk' (*risque* in French), rather than a sustained framing. Indeed, the word 'risk' was found 15 times throughout the articles. However, it was not particularly salient (6 per cent of the articles), nor was it a dominant tone (five articles). Indeed, it was usually combined with the implicit risk frame when it was a dominant tone.
- The **opportunity** frame was nearly absent (just two articles out of 34), and was neither salient nor dominant.
- The **IPCC concepts** of likelihood and probability were present in 11 of the 34 articles (32 per cent), and in more than half of the cases the concepts were explained (six articles).

Coverage of the Arctic sea ice melt:

- The **uncertainty** frame was present in 19 of the 25 articles coded (76 per cent), but a significant number of these was due to the presence of the 'when will the Arctic be ice-free?' question (12 articles). It was not particularly salient, but nearly half of the articles had it as a dominant tone (ten articles – 40 per cent) and had uncertainty quotes within them (11 articles – 44 per cent). It should be noted that only two articles mentioned sceptics.
- The **implicit risk** or disaster frame was particularly present since it was found in 20 of the 25 articles (80 per cent). It was also significantly more salient than the uncertainty frame with 12 articles (48 per cent) having it as a headline or dominant in the opening few lines. Despite its presence in 80 per cent of the articles, the implicit risk constituted the dominant tone of an article for only 40 per cent (ten articles).

However, it was particularly found in the linguistic repertoire of adjectives such as 'dramatique', 'dangereux', and 'catastrophique' which were found in 44 per cent of the articles.

- The **explicit risk** was less present although it was found in a third of the articles (eight out of 25). It was, however, not really salient (only in two articles). Six articles (24 per cent) had it as a dominant tone but this was always in conjunction with other frames (uncertainty/ implicit risk/opportunity).

- The **opportunity** frame was present in ten articles out of 25 (40 per cent). It was salient in four of them, and a dominant tone in six (although in five of these articles this was used in conjunction with other tones). Four articles included direct quotes stressing opportunities (16 per cent). Although the word 'opportunity' itself appeared only once, the opportunity frame was much more present in the coverage of the Arctic than in the coverage of the three IPCC reports (6 per cent of the articles in the case of the IPCC against 40 per cent of the articles in the case of the Arctic).

- The **specific possible impacts** mentioned the most were the effects on the local ecosystem, including polar bears, the changes in air and sea temperatures in the Arctic, and heightened political tensions (found in five articles each), followed by possible colder winters in the northern hemisphere and the effects on the ecosystem from mineral, gas, and oil exploration (four articles each). A changing local climate in the Arctic and the effect of the sea ice melt on lives of local people were possible impact found in only three articles each. The least mentioned impact was possible methane release (two articles). Regarding the possible opportunities, improved chances of exploration were mentioned in eight articles and new shipping routes in seven.

- Five of the articles included direct **quotes from scientists expressing emotion** at the state of the ice melt, which were surprise (two), worry (two) or shock (one).

As regards overall trends, the **uncertainty frame** was present in a high proportion of the articles (81 per cent). Nearly half of the articles contained direct quotes stressing uncertainty (41 per cent) but it was not particularly salient (17 per cent). One article out of five mentioned sceptics but their views mainly represented a marginal portion of the articles, with most of the articles stressing where the scientific consensus lies (only two articles out of 12 did not make clear where the consensus lies). An interesting finding was that mention of sceptical voices was found only in broadsheet newspapers, and mainly in *Le Monde* (nine articles) as opposed to *Le Figaro* (three) which traditionally offers more room for sceptical voices. It should be noted, however, that this can be explained by the fact that there were significantly more articles from *Le Monde* in the analysis (29 articles) than from *Le Figaro* (20 articles). And, as highlighted previously, sceptics' views represented a marginal portion of the articles at odds with the scientific consensus (one of the two articles which did not make clear the consensus was an opinion piece by a sceptic). Finally, the uncertainty frame was a dominant tone in roughly one third of the articles.

The **implicit risk** frame was equally present in a high proportion of the articles (81 per cent). However, it was significantly more salient than the uncertainty frame (58 per cent). More than half of the articles used quotes stressing implicit risk and nearly half of the articles contained strong adjectives stressing this implicit risk frame. It was also the most common dominant tone, since it was found in 53 per cent of the articles. Finally, more than one article out of five contained metaphors that expressed implicit risk. The most common one was the comparison of the climate to a 'machine that could spin out of control', or an 'engine racing and becoming uncontrollable'. Such a metaphor objectifies the planet as a complex system/mechanism that needs controlling/fixing. Another interesting metaphor in the coverage of the Arctic sea ice melt referred to the increasing political tension by comparing it to the board game 'Battleship', hence stressing the political issues at stake.

The **explicit risk** frame was much less present but was nonetheless found in 37 per cent of the articles. It was, however, rarely salient (7 per cent). Although it constituted a dominant tone in 19 per cent of the articles, it was almost always so in conjunction with other tones (ten articles out of 11). It was a dominant tone in five *Le Monde* articles, five *Le Parisien* articles, and only one *Le Figaro* article. In period 1, the explicit risk tone was combined mainly with the implicit risk tone (two out of two). This was the case for instance in *Le Figaro* which in January 2007 published an opinion piece by Hervé Le Treut, one of the most quoted scientists throughout the newspapers. He stressed that given the fact that 'we are facing a risk whose general contour we are grasping', politicians need to take note of the scientific results and act accordingly. Le Treut argued that it was now necessary for politicians and citizens alike to take action with a target of halving GHG emissions by 2050, as otherwise the risks from temperature rises may have a devastating impact.

In period 2, explicit risk was again combined with the implicit risk frame. However, the explicit risk frame was particularly dominant in a long article in *Le Monde* (2007) which dealt with great rivers being under threat, and how this related to what the IPCC identified as the main concern resulting from global warming, water supply. It included the use of a strong adjective to stress the existing risks ('the greatest risk'/'le plus fort risque'). In period 3, there was only one article with explicit risk as a dominant frame and it also included uncertainty as a dominant tone. Finally, in period 4, the explicit risk tone was combined in various ways with the three other tones. For example, an article in *Le Monde* combined the explicit risk tone with both the uncertainty and opportunity tones (Truc, 2011). Following *Le Monde*'s broadsheet agenda, the article dealt with the reconfiguration of geopolitical and economic interests in the region, and provided a good example of how opportunity, uncertainty, and risk intertwine at the crossroads of scientific research and political decision making.

Finally, the **opportunity** frame was the least present since it was found in only 20 per cent of the articles. It was neither salient (7 per cent) nor a dominant tone (10 per cent – six articles, five out of which used a combination of dominant tones). It should also be noted that this opportunity frame was virtually absent from the coverage of the IPCC reports and mainly found in the coverage of the Arctic sea ice melt.

India
by Anu Jogesh and James Painter

Climate change has been described as a 'non-issue in Indian organised politics' (Navroz, 2013). However, from time to time it does feature in political discourse. For example, debates took place in the upper and lower house of parliament before and after the 2009 Copenhagen summit, when the topic gained enough momentum to elicit reactions from politicians. This trend has continued around subsequent UN summits. While the seriousness of climate change and its potential risks are unquestioned, political debates focus less on addressing the issue and more on concerns that India might lose ground by committing to legally binding action which could limit its economic development. Speeches given by the Opposition in the Rajya Sabha (the upper house) during the UN's 2010 climate conference in Cancun are typical of this sentiment (Jaitely, 2010).

India is a heavy user of coal power plants, and the world's fourth largest emitter of GHGs in absolute terms (but much lower than that in per capita terms). This is one of the reasons it is now seen as a key participant and an important determinant of outcomes in international meetings on cutting emissions. For example, at the UN's Durban conference in December 2011, the Indian delegation was a crucial player in a last-minute agreement that committed all countries – developing and developed – to agreeing to a deal on cutting carbon emissions by 2015. In 2008 it announced several voluntary measures, including the enhanced use of solar energy, with the aim of cutting its emissions intensity from business as usual levels by 2020 (Government of India, 2010).

In sharp contrast to the UK, Australia, and the USA, climate scepticism has never publicly entered Indian political debates. In fact, a headline from *The Times of India* in December 2009 reads, 'No climate sceptics in Lok Sabha' (the lower house) (*The Times of India*, 2009). The article quotes various MPs, including those from the Opposition, acknowledging the seriousness of the problem, but not questioning any aspects of the science. It is significant that a word search on climate change on the government's Press Information Bureau website reveals discussions and responses on India's stand in international conferences, aspects of its national and sub-national policy on climate change, and details about its national communication reports, as well as political concerns about regional and sectoral climate vulnerability and adaptation. But climate scepticism does not get a mention.

Recent opinion surveys of Indians would suggest that, while there is broad agreement on observed environmental changes, knowledge and understanding of these changes through the prism of climate change is low.[6] In 2007, the BBC World Service conducted a poll among 22,000 people in 21 countries on their understanding and beliefs towards climate change (BBC World Service, 2007). Forty-seven per cent of the Indians surveyed stated that human activity was the principal cause of climate change, while 21 per cent believed otherwise. However, 33 per cent did not know or did not provide an answer.

In 2012 the Yale Project on Climate Change and Communication carried out a survey on climate change among 4,013 Indians (Leiserowitz and Thaker, 2012). The study elicited a wide response on the observation of local environmental changes. For instance 80 per cent of respondents spoke of changes in rainfall patterns in the past decade. In terms of awareness, only 7 per cent of the sample stated that they knew 'a lot' about global warming.

On offering individuals a short description of climate change, 72 per cent said they believed it was happening, half of the sample said they had felt its effects, and 56 per cent blamed it on human activities. This suggests, as stated above, that while many Indians are taking note of environmental and climatic changes, they do not always associate them with the idea of global warming. It is a view shared by Meena Menon, the Chief of Bureau for *The Hindu* in Mumbai, who says, 'In Ladakh, Anantpur, even parts of Mumbai, people have noted various climatic changes; erratic rainfall, retreat in glaciers, coastal erosion, and so on. The science has always followed later.'[7]

The Yale University study also revealed that in terms of information on global warming, scientists were the most trusted source (73 per cent), followed by the media (69 per cent), then environmental organisations (68 per cent) and finally the government (50 per cent). In terms of media consumption, 65 per cent of the respondents of the survey watched television, and 54 per cent said they read newspapers.

In fact, the Indian media scene is thriving. There are around 180 news channels and by the time of the next general election, scheduled for 2014, the number is expected to rise to well over 250 (Roy, 2012). In contrast to the declining circulations and revenues in much of the Western world, the Indian press is also booming. *The Times of India* (*TOI*), a 150-year-old English-language newspaper owned and published by Bennett, Coleman & Co., has the largest daily readership among English dailies at over 7.6 million in 2012, with a circulation of 1.2 million in Delhi alone (MRUC and Hansa Research, 2012). Indeed, the Audit Bureau of Circulation describes *TOI* as the largest English-language daily in the world. It has a pan Indian presence with regional centres in 26 Indian cities.

In an October 2012 article the *New Yorker* highlighted the paper's focus on shorter articles, 'snappier' sentences, 'more sports, less politics, more Bollywood, more color' (Auletta, 2012). The author of the article talks of how '*The Times of India* sees itself not as an agenda-setter but as a bulletin board'. *The Hindu* projects itself as an antithesis to the image of the *TOI*. Whereas the *TOI* is keen to cater to a broad base of English-speaking readers, *The Hindu* in its official website talks of how its coverage and editorial stand has 'won for it the serious attention and regard of the people who matter'.[8] *The Hindu* is the third largest English-language Indian daily by readership with over 2.2 million readers. It is based in the southern metropolitan city of Chennai and run by Kasturi and Sons and, while it too enjoys a pan Indian presence, it has a loyal readership in the south of the country. It is widely regarded as a left-leaning paper.

Compared to the *TOI* and *The Hindu*, the *Business Standard* is a relatively younger and smaller business and financial daily. The paper, headquartered in Mumbai, is run by an independent board of directors, although the Kotak Mahindra Group, a well-known financial institution, has a majority stake in it. The paper was first launched in 1975 and currently enjoys a quarterly readership of 1.5 million copies. It is the third largest business daily in the country. What distinguishes it from other national and financial dailies is the variety of editorial contributors, who range from well-known economists, political thinkers and academics to policy makers. The paper's readership, though relatively small, is quite targeted, particularly towards the business and policy elite in its columns.

In general, climate news coverage has become a regular feature in English-language media in India. On any given day, a cursory search for climate-based news across all mainstream Indian print media will reveal an average of five articles every day, not including wire reports or online news sites. Max Boykoff's chart of the volume of all climate change stories in the English-language Indian print media coverage (which includes the *TOI*, *The Hindu*, the *Indian Express*, and the *Hindustan Times*) suggests that it peaked in 2007, and then again around the time of the summits in Copenhagen and Cancun, followed by a fairly steady 150 articles a month since then.[9]

The frequency and consistency of climate-based reports are driven by two main factors. The first are events, and specifically (though not limited to) UN climate conferences known as COPs. Apart from the participation in such conferences of Indian negotiators, the business community and civil society organisations, there is political and news interest in how India fared in maintaining its equity position and not crossing its 'red lines' of sacrificing development for emission cuts. The coverage around COPs has been bolstered by Indian and international NGOs funding reporters to travel and cover them.

The second is the presence or absence of dedicated climate or environment reporters. While non-climate reporters are often roped in during important climate-related events, the regularity of in-house climate coverage is directly related to the presence of specialist reporters. For instance, the frequency of climate news in the business daily *Mint* dropped when its reporter took another position in the organisation in early 2011. The *Indian Express* saw a relative dip in its in-house climate coverage after 2008 when its science and environment correspondent left. Indeed, some observers note the lack of investment in the Indian newspapers' own reporting. The *TOI* is one of the few newspapers to have a dedicated reporter tracking climate news since 2007, which has resulted in a stream of in-house articles about climate change, though they largely deal with the politics around it.

There are sharp differences in the source, frequency, and relative importance given to political as opposed to science-based climate news. Compared to political pieces, science news stories are relatively fewer in number. The latter are predominantly based on international wires or other international sources, unless they are linked to an Indian research finding or a specific weather event.

Climate scepticism rarely appears in the Indian print media. A previous RISJ study, *Poles Apart*, found that the number of articles in the Indian media mentioning or quoting climate sceptics was very low compared to the number quoted in the USA and UK (Painter, 2011: 59). Unlike the media in these two countries, no Indian newspaper or media group has a conspicuous editorial stand on accepting or denying climate change. The media and civil society organisations which shape the discourse on climate change in India are more concerned with issues of vulnerability, adaptation, and maintaining equity in the mitigation debate.

The number of studies which examine climate discourse in the Indian media is relatively few when compared to similar studies in industrialised nations. An oft-cited paper by Simon Billet looked at how climate change was represented in four English-language dailies between 2002 and 2007 (Billet, 2009). The paper found that the news typically focused on India's vulnerability to climate risks but placed the responsibility for taking action on industrialised nations. The study also highlighted the fact that the Indian press entirely endorses climate change as a scientific reality.

A more recent study by Anu Jogesh examined Indian print media coverage of climate change in the seven months surrounding the Copenhagen summit in December 2009 (Jogesh, 2011). Her results, based on nine national and financial dailies, reveal that with a sharp increase in reporting around the conference, there was a broadening in the discourse on climate change. While the dominant rhetoric for industrialised nations to take the lead on climate action persisted, there was a growing belief that India could do more. Among other studies (Aram, 2011; Boykoff, 2010; Painter, 2010: 137) is one by Reusswig and Meyer-Ohlendorf (2012), which focused on climate impacts on Indian megacities. The study examined two English-language and two Telugu papers and made similar observations to the Jogesh paper, stating that climate discourse in the recent past had become more differentiated, conceding some degree of responsibility for climate action to countries like India.

It is clear from the above discussion why the three newspapers were included in this study. *TOI* is one of the only papers to have a dedicated climate reporter, and publishes articles on the issue regularly. However, science-based news is often relegated to the international pages and is almost entirely wire driven. *The Hindu* tends to carry events-based news as well as agency articles related to climate change, and on occasions focuses on climate impacts and adaptation especially in the agriculture sector. Science news is also predominantly sourced from wire reports and international articles. Finally, while the *Business Standard* has a relatively limited readership, its audience is targeted. Science-based news in the paper is sporadic, but opinion pieces on climate change are written by, and for, those who regularly research or dictate climate policy.

The Indian print media's coverage of risk and uncertainty

A summary of the results of the coding of 56 articles in the three newspapers can be seen in Table 6.3. For the coverage of the three IPCC reports, the results show that:

- Of the 19 articles across the three periods, 68 per cent (13 articles) mention **uncertainties** surrounding climate science, including future projections, and 37 per cent (7 articles) offer specific ranges of uncertainty. Only 26 per cent (5 articles) use words such as 'increasing certainty' or 'almost certain'. Often uncertainty does not necessarily pertain to temperature projections but to the uncertainty of climate risks and impacts. For instance, a piece in the *Business Standard* states that 'Nobody knows how many glaciers there are on the Indian side of the Himalayas. The rapid speed at which these glaciers are melting may well change the basic economics' (Menon, 2007). This straddles the uncertainty and the implicit risk frame.
- While the uncertainty frame is relatively present, the **implicit risk** frame is clearly the predominant one: 18 out of 19 articles (95 per cent) offer sourced statements mentioning implicit risks, and the same high percentage has it as the dominant tone. It is salient in 42 per cent. As many as eight articles have headlines of implicit risk. And nearly 90 per cent contain a quote from a scientist or scientific report containing the same frame.

Table 6.3 India Results

	Period 1 WG-1	Period 2 WG-2	Period 3 SREX	Total IPCC	%	Period 4 Arctic	%	Total	%
Number of articles	8	6	5	19		37		56	
Uncertainty	**6**	**4**	**3**	**13**	**68**	**23**	**62**	**36**	**64**
More certainty	2	0	3	5	26	14*	38	n/a	n/a
Duelling experts	2	0	0	2	11	n/a	n/a	n/a	n/a
Salience	3	0	1	4	21	5	14	9	16
Direct quotes	6	5	3	14	74	16	43	30	54
Word presence	0	0	4	4	n/a	2	n/a	6	n/a
Sceptics	0	0	0	0	0	1	3	1	2
Dominant tone	3	0	1	4	21	11	31	15	27
Implicit risk	**8**	**6**	**4**	**18**	**95**	**28**	**76**	**46**	**82**
Salience	3	3	2	8	42	9	24	17	30
Direct quotes	7	6	4	17	89	16	43	33	59
Adjectives	6	5	2	13	68	11	30	24	43
Metaphors	2	0	0	2	11	11	30	13	23
Dominant tone	8	6	4	18	95	18	49	36	64
Explicit risk	**0**	**5**	**2**	**7**	**39**	**2**	**5**	**9**	**16**
Salience	0	0	0	0	0	0	0	0	0
Direct quotes	1	1	2	4	21	1	3	5	9
Adjectives	0	1	0	1	5	0	0	1	2
Metaphors	0	0	0	0	0	0	0	0	0
Word presence	1	7	8	16	n/a	2	5	18	n/a
Dominant tone	2	3	2	7	39	0	5	7	13
Opportunity	**1**	**2**	**0**	**3**	**16**	**9**	**24**	**12**	**21**
Salience	0	0	0	0	0	6	16	6	11
Direct quotes	0	1	0	1	5	2	5	3	5
Word presence	0	0	0	0	n/a	0	0	n/a	n/a
Dominant tone	0	0	0	0	0	8	22	8	14
IPCC concepts	**0**	**4**	**3**	**7**	**37**				
Explanation	0	0	1	1	5				

* Year ice-free

- The content of the reports indicates nine articles that predominantly focus on societal impacts as opposed to seven that delineate scientific impacts. This corroborates previous studies that have highlighted the Indian media's focus on climate impacts rather than mere temperature projections, as dangers are construed as much closer to home.
- The **explicit risk** frame is reasonably present (seven out of 19 articles) but it is never salient. The word 'risk' is common, with 16 uses across the period. It is present as a dominant tone in seven articles, but always in combination with implicit risk.
- The **opportunity** frame only appears three times; two of these are in the *Business Standard* and both are sourced from the *Financial Times* in London. The first is a business piece which quotes Lehman Brothers as stating, 'global capital goods companies, which supply plant and equipment to the power generation industry, are likely to be net beneficiaries'. The second, which is headlined 'the worst damage will occur in the developing world' uses the IPCC report to predict that some regions in the UK, northern Europe and the USA will benefit from climate change – in other words the opportunities from not doing anything to tackle GHG emissions.
- While seven of the 19 articles offer **IPCC concepts** of likelihoods and the same number of pieces ascribe it to the IPCC, only one explains what the IPCC means by these concepts.

Arctic sea ice melt:

- Twenty-three out of 37 Arctic articles (62 per cent) have a presence of the **uncertainty** frame, of which 14 mention the doubts about which year the Arctic will be free of ice. Over 40 per cent carry quotes from scientists or reports mentioning uncertainties, and over 30 per cent have uncertainty as the dominant tone.
- As with the IPCC reports, the **implicit risk** frame is the most dominant of all four frames (in 76 per cent of the articles). Twenty articles (54 per cent) carry general unsourced statements of implicit risk and 16 (43 per cent) have sourced statements from scientists, experts, or scientific reports mentioning the adverse impacts. Moreover, it is the dominant tone in nearly half of the articles.
- While the uncertainty and implicit risk frames obviously dominate, as many as 24 per cent (nine articles) mention climate-related opportunities. The **opportunity** frame, while not being the leading

frame, was present in significant numbers in the articles pertaining to Arctic ice melt as compared to three pieces on the IPCC reports. Predictably, most of them point to the further opening of shipping routes and the chance to exploit mineral resources.

- Almost all the articles with the opportunity frame were written in 2011 and 2012. Only one piece was penned in 2010, perhaps pointing to the fact that this is a more recent development. All but four of the reports belong to the *Business Standard* and *The Hindu*, and two are editorials linking this opportunity to the politics of climate change between industrialised and industrialising nations.
- There are only two articles that have a presence of **explicit risk**, confirming that this is not a dominant frame across all four periods.

General results

Across all 56 articles included in this survey, the **implicit risk** frame was by far the most present (in 82 per cent), the most salient (30 per cent), and most frequently the dominant tone (64 per cent). Moreover, nearly 60 per cent had direct quotes about the adverse impacts within them, and adjectives such as 'severe', 'grim' and 'devastating' were found in 43 per cent. Indeed, this frame figured as the highest in each category monitored.

The **uncertainty** frame came next with a presence in over 60 per cent of the articles. Over half the articles also contained quotes which implied uncertainty. It was a dominant tone in about a third. Prior studies have shown that climate scepticism is all but absent in the Indian media. This study corroborates this. There was only one article, sourced from Associated Press, in the entire sample that contained a sceptical voice, and even then the dominant consensus in the piece was that anthropogenic climate change is a reality.

The **explicit risk** frame came last. In all nine articles where it was a dominant tone (16 per cent of the total), it was always present in combination with the implicit risk frame. For example, a report in *The Hindu* mentions the word risk and also states that 'there is very high confidence that many natural systems are being affected by regional climate changes'. While this is indicative of the explicit frame, the piece is also peppered with generic implicit risk framed statements such as 'the world will face heightened threats of flooding, severe storms and the erosion of coastlines' (*The Hindu*, 2007).

In similar fashion, an article in the *TOI* of April 2007 contained both a statement that 'the entire coastline will face up to 20 per cent increased risk of cyclonic storms', and at the same time a quote from the Indian chairman of the IPCC, Rajendra Pachauri, saying that 'India needs to worry about food security, the extreme heat waves that are going to hit it, changes in precipitation, water scarcity and glacial melting. The impacts are now clear and better understood than ever before' (Sethi, 2007). An article in *TOI* (2012) based on Associated Press from period 3 which uses the explicit risk frame and language was clearly prompted by the risk language used in the SREX report.

Finally, it is worth stressing that, as stated earlier, the Indian print media are predominantly focused on the politics of climate change, at least in terms of in-house reportage. Science-based pieces often take a back seat. Not surprisingly then, 40 out of 56 articles were international pieces, which means they are either international wire reports or sourced from international dailies, such as the *Guardian*, the *New York Times* or the *Financial Times*. Only ten of the articles in the sample were written by in-house journalists.

Norway
by Christian Bjørnæs and Anja Naper

In April 1987, the Brundtland Commission, chaired by then prime minister of Norway Gro Harlem Brundtland, published the UN Commission report *Our Common Future*, which both warned of a warming climate and launched the concept of sustainable development. The report provided the momentum for the 1992 Earth Summit in Rio de Janeiro and was instrumental in creating a domestic image of Norway as an ethical leader in the fight against climate change (Eide and Ytterstad, 2011).

In the early 1990s Norway adopted several measures to limit GHG emissions, among them the carbon tax of 1991 which provided an incentive to stop flaring on offshore installations and inject the carbon dioxide underground. The extraction and refinement of Norway's offshore petroleum resources make up around 29 per cent of the country's carbon emissions (Statistics Norway, 2013). In the run-up to the UNFCCC's meeting in Kyoto in 1997, the Norwegian government was instrumental in promoting flexible mechanisms which would allow Norway to offset domestic carbon emissions by financing emission reductions in other countries (Asdal, 2012). These mechanisms would enable Norway to continue its petroleum activities while fulfilling its international obligations to reduce GHGs. Under the Kyoto Protocol, Norway agreed to stabilise its emissions at 1990 levels and has achieved this goal without the use of carbon trading (Government of Norway, 2013).

In the late 1990s and early 2000s, Norway did not introduce any new measures to reduce GHG emissions. However, in 2007 and 2008, Prime Minister Jens Stoltenberg gave two landmark speeches: at the Labour party's conference in May he promised to cut Norway's GHG emissions by 10 per cent by 2012 and by 30 per cent by 2020. In 2050 Norway would become carbon neutral which would be achieved partly by reducing domestic emissions and partly by carbon offsetting. On 1 January 2008, he announced his plans to make Norway a leader in carbon capture and storage, a technology yet to be developed that would trap and store carbon dioxide underground.

The emission reduction targets were made official Norwegian policy in a parliamentary agreement called the Climate settlement of 2008. The agreement was signed by all major parties except the far-right Progressive party and stated that two thirds of the emission reductions would be achieved domestically.

From 2007 to December 2009, two technological issues were salient in the Norwegian public debate: carbon capture and storage, and Norway's status as a leading producer of renewable energy. The failure of the Copenhagen summit in 2009 had a profound effect on the Norwegian public debate. In 2011 and 2012, several new oil and gas fields were discovered in Norwegian waters. Optimism in renewable technologies was replaced by renewed petroleum optimism.

Since 1985 Norsk Monitor, a poll run by Ipsos MMI, has measured Norwegians' values and attitudes to a range of issues. It clearly shows that attitudes to environmental protection change with the population's values. In periods when Norwegians' values are more idealistic and less materialistic, their willingness to give priority to environmental protection is higher. In the late 1980s Norwegians were relatively idealistic but throughout the 1990s they became gradually more materialistic. Between 2003 and 2007 there was a sudden shift towards more idealistic values but in 2009 there was a noticeable shift back towards materialism. Concerns about climate change follow the same pattern. It was relatively high in the late 1980s and fell gradually through the 1990s, peaked in 2007 and dropped in 2009 (Hellevik, 2010).

A Synovate poll from 2008 showed that 32 per cent of Norwegians doubted that climate change was anthropogenic. This proportion increased to 39 per cent in 2010 (Lindsay Griffin and Campbell Lehne, 2010). Although the question was phrased differently, a poll from TNS Gallup confirmed the pattern. The proportion that dismissed anthropogenic contribution increased from 15 per cent in 2009 to 20 per cent in 2010. Despite the fact that 80 per cent accepted the anthropogenic contribution to climate change, 70 per cent supported increased petroleum production even though it would increase emissions of GHGs, according to the same poll (Rees, 2012). One reason could be the low perception of threat. According to a Gallup survey of global opinions on climate change in 2007–8, only 43 per cent of Norwegians perceived climate change as a threat. In neighbouring Sweden, the figure was 56 per cent, while 63 per cent of Americans, 69 per cent of British, and 75 per cent of Australians responded that global warming was a serious personal threat (Pugliese and Ray, 2009).

Unlike in the US and UK, Norwegian climate sceptics are not well organised and there have been no proven ties to think tanks, secret funds or the fossil fuel industry. The organisation 'Klimarealistene' (The climate realists) was established in May 2008 and consists of roughly 800 members. Their leader, Ole Henrik Ellestad, a professor in chemistry and former deputy director at The Research Council of Norway, is perhaps the country's most prominent climate sceptic. Norwegian sceptics typically dismiss the concept of anthropogenic warming arguing that natural variations are driving the climate or, as more recently, they dismiss the warming trend altogether.

Several studies have investigated how the Norwegian media report climate change. Marianne Ryghaug argues in her 2006 study of eight Norwegian newspapers, called 'Some like it hot', that the main focus in their reporting is climate catastrophe, uncertainties, and conflicts between individuals, groups or institutions as a way of capturing the readers' attention (Ryghaug, 2006). In the journalists' quest for balance, they use a wide range of sources including marginalised groups such as climate sceptics, according to Ryghaug. In a later study comparing the Norwegian and Swedish coverage, Ryghaug found that 'the drama of scientific disagreement is mostly a Norwegian phenomenon. Swedish newspapers often emphasised scientific consensus rather than disagreement' (Ryghaug and Skjølsvold, 2009).

Katherine Duarte investigated which attitudes and voices were represented in the Norwegian coverage of climate change from October

2007 to April 2008. She found less disagreement than expected: 73 per cent of the articles analysed agreed with the IPCC consensus on anthropogenic climate change, 20 per cent were neutral and only 8 per cent were critical (Duarte, 2010). The highest representation of critical voices, 11 per cent, was found in *Aftenposten,* which is the paper which gives most attention to Norwegian climate sceptics. In a later analysis of five papers spanning the autumn of 2009 to the autumn of 2011, Duarte concludes that sceptical voices were somewhat visible right after 'Climategate' and during the Copenhagen summit in 2009. However, during the winter of 2010, these voices decreased. 'Klimarealistene' were highly visible in the period under investigation and account for 11 per cent (Duarte, 2012).

Perhaps the most pronounced trait of Norwegian climate change reporting is the focus on domestic issues and its reliance on domestic sources. In Duarte's dissertation, '66% of the articles had a national focus'. In a comparative study of 12 countries' reporting from the Bali climate summit in 2007, Norway ranked top in the use of domestic sources: over 80 per cent of the sources quoted were Norwegian. In Sweden and the US, domestic sources made up about half the reporting (Eide and Ytterstad, 2011). In a later study by Eide of the reporting from the Copenhagen summit in 2009, a little over half of the sources in the Norwegian reporting are domestic. This is still substantially higher than the average of 40 per cent in the 19 countries investigated (Eide, 2011).

Aftenposten is a high quality national daily newspaper. Until 20 December 2012 it had a morning and evening edition. The latter was a local newspaper for the Oslo metropolitan area. It is the largest paper by circulation with an average daily distribution of 322,000 copies, most of it by subscription.[10] It was traditionally a right-leaning paper but its editorial line is less ideological today than it was a few years ago. The paper is owned by Schibsted, Scandinavia's largest media group. It was the only Norwegian newspaper with a full-time environmental correspondent, Ole Mathismoen, from 1997 until he was promoted to head of the political desk in 2010.

VG is a national daily tabloid and, with its 690,000 average daily readers in 2012, it is the country's largest newspaper by readership. From 1981 to 2010 it was also the largest by circulation. It is also owned by Schibsted and is considered a classic tabloid with no pronounced political leaning. The newspaper does not have an environmental correspondent.

Finally, *Dagbladet* is a national daily tabloid and Norway's third largest newspaper with an average circulation in 2012 of 88,500 copies. It is owned by the private company Berner Gruppen. The largest shareholder is Jens P. Heyerdahl who holds effective control of the newspaper through several different companies. *Dagbladet* is traditionally a moderate left-wing paper.

The Norwegian media's coverage of risk and uncertainty

The results of the coding can be seen in Table 6.4. The results for the coverage of the three IPCC reports would suggest the following:

- The **uncertainty** frame was present in 19 of the 22 articles, or 86 per cent of the sample. However, this frame was only salient in two articles (9 per cent). Both articles appeared in period 1. Uncertainty was the dominant tone in only four articles (18 per cent) of which three appeared in period 1 and one in period 3.

  Eight of the 14 articles where the uncertainty frame was present contained such terms as 'increasing certainty', 'almost certain' or 'most certain to date', when referring to climate science, including future projections. Direct quotes about uncertainty were found in six articles.

  Of the articles where the uncertainty frame was present, 11 also contained the implicit risk frame. Seven of these articles were from period 1, three were from period 2 and one from period 3.

  Sceptical comment was only present in three articles, or 14 per cent of the sample. In one article, a wire story which appeared in *VG*, the sceptical voices made up half or more. In two *Aftenposten* articles, the sceptical voices made up less than 50 per cent and 10 per cent respectively. All three articles gave context about the scientific consensus.
- The **implicit risk** or disaster frame was just as prominent as the uncertainty frame in Norway (19 of 22 articles, or 86 per cent). Implicit risk was salient in 13 articles or 59 per cent of the sample. This was the most frequently used frame in the Norwegian reporting. It was the dominant tone in 14 articles (64 per cent).

  Strong adjectives such as 'enormous', 'large' and 'substantial' were used in 15 articles (68 per cent) to stress implicit risk. Direct quotes, on the other hand, were only used in seven articles (32 per cent).
- The **explicit risk** frame was largely absent. It was present in three articles (14 per cent) and salient in none. It was the dominant tone in two articles alongside implicit risk. Seven articles (32 per cent)

Table 6.4   Norway Results

	Period 1 WG-1	Period 2 WG-2	Period 3 SREX	Total IPCC	%	Period 4 Arctic	%	Total	%
Number of articles	16	3	3	22		21		43	
Uncertainty	**14**	**3**	**2**	**19**	**86**	**16**	**76**	**35**	**81**
More certainty	7	0	1	8	36	7*	29	n/a	n/a
Duelling experts	3	0	0	3	14	n/a	n/a	n/a	n/a
Salience	2	0	0	2	9	3	14	5	12
Direct quotes	5	1	0	6	27	7	33	13	30
Word presence	11	0	2	13	n/a	0	0	13	n/a
Sceptics	3	0	0	3	14	1	5	4	9
Dominant tone	3	0	1	4	18	3	14	7	16
Implicit risk	**13**	**3**	**3**	**19**	**86**	**13**	**62**	**32**	**74**
Salience	8	3	2	13	59	3	14	16	37
Direct quotes	5	1	1	7	32	13	62	20	47
Adjectives	10	2	3	15	68	4	19	19	44
Metaphors	0	0	0	0	0	0	0	0	0
Dominant tone	9	3	2	14	64	10	48	24	56
Explicit risk	**1**	**2**	**0**	**3**	**14**	**1**	**5**	**4**	**9**
Salience	0	0	0	0	0	0	0	0	0
Direct quotes	5	1	1	7	32	0	0	7	16
Adjectives	0	0	0	0	0	4	19	4	9
Metaphors	2	0	0	2	9	0	0	2	5
Word presence	2	2	0	4	n/a	1	n/a	5	n/a
Dominant tone	1	1	0	2	9	0	0	2	5
Opportunity	**3**	**1**	**0**	**4**	**18**	**10**	**48**	**14**	**33**
Salience	0	0	0	0	0	4	19	4	9
Direct quotes	0	0	0	0	0	2	10	2	5
Word presence	0	0	0	0	n/a	6	n/a	6	n/a
Dominant tone	0	0	0	0	0	7	33	7	16
IPCC concepts	**10**	**0**	**1**	**11**	**50**				
Explanation	4	0	0	4	18				

* Year ice-free

contained direct quotes about explicit risk by named scientists or scientific reports. The word risk appeared in four articles (18 per cent).

- The **opportunity** frame was also largely absent. It featured in four articles (18 per cent), was never salient and never the dominant tone. The word opportunity was never used and no scientist or scientific report pointed out opportunities posted by climate change.

- The **IPCC concepts** of likelihood and probability were present in 11 articles (50 per cent), but were explained in only four (18 per cent).

Results from the coverage of the Arctic sea ice melt were as follows:

- The **uncertainty** frame was present in 16 articles (76 per cent). It was salient in three articles and the dominant tone in three articles (14 per cent). Seven articles (33 per cent) contained direct quotes from named scientists or scientific reports about uncertainties. Eleven of the articles containing the uncertainty frame also contained the implicit risk frame. Seven of the uncertainty articles also contained the opportunity frame.

- The **implicit risk** frame was present in 13 articles (64 per cent). Although it was salient in only three articles (14 per cent), it was the dominant tone in ten (48 per cent). Implicit risk was evident in quotes from scientists or scientific reports in 11 articles (52 per cent). It was rarely communicated through the use of adjectives (four articles, 19 per cent) and never through metaphors. Of the 13 articles containing the implicit risk frame, 11 also contained the uncertainty frame.

- The **explicit risk** frame was only present in one article (5 per cent) and it was never salient nor the dominant tone.

- The **opportunity** frame was present in ten articles (48 per cent). Of these articles, seven also contained the uncertainty frame. The opportunity frame was salient in four articles (19 per cent) and dominant in seven articles (33 per cent). The word opportunity was used in four articles (33 per cent).

- In two of the 21 articles scientists expressed worry.

Analysis of results

Two periods stand out as heavily reported: the launch of the IPCC's WG-1 report of February 2007, and the melting of the Arctic sea ice. These two periods make up 86 per cent of the sample (37 articles). The launch of the WG-2 report and the SREX received little attention, with only three articles each. A range of factors could explain this: in Norway, there was a coordinated media and communications strategy ahead of the launch of the WG-1 report. The Climate and Pollution Agency, the IPCC's Norwegian focal point, in collaboration with key research institutes, had produced fact sheets for journalists and organised a press conference, attended by approximately 200 journalists, which ran simultaneously to the IPCC's own launch on 2 February. The WG-2 report was launched on 6 April 2007, which was Good Friday that year. There was no press conference and no back-up material was provided to help the journalists understand the science. Much of Norway closes down during Easter and although there were newspapers coming out the following Saturday, most of the content was prepared in advance. It is not clear why the SREX report received such little attention, but it is consistent with other country results. The frequent reporting on the Arctic sea ice is probably due to its proximity to Norway and the substantial economic interests Norway has in this area.

Reviewing all the articles, it is clear that there is a very strong presence of uncertainty and implicit risk (81 per cent and 74 per cent respectively). A significant amount of uncertainty in the reporting of the IPCC reports is due to mentions of uncertainties about climate science, including future projections. This was included in all 19 articles containing uncertainty. Fifteen articles contained uncertainty parameters or ranges in projections, while only two articles mentioned shortcomings of computer models. Only three articles featured duelling experts suggesting contestation about some aspect of the science.

Looking at the reporting of the Arctic, all 16 articles containing uncertainty mentioned uncertainties about the future impact of sea ice melt. Other uncertainties were not as prominent. Seven articles included uncertainty about when the Arctic might be free of ice in the summer, and four included uncertainty about possible impact of the albedo effect on hastening the ice melt. Other uncertainties mentioned were the impacts sea ice melt could have on weather at lower latitudes and on polar bears.

The implicit risk frame was just as dominant as the uncertainty frame in the reporting of the IPCC reports (86 per cent), but was much less prominent in the reporting of Arctic sea ice melt (62 per cent).

Implicit risk found its way into 17 of the 19 IPCC articles using this frame through named sourced statements from scientists/experts or scientific reports mentioning risks. General, unsourced statements mentioning implicit risks were found in 13 articles. A majority of the articles stressed the societal impacts over the scientific impacts from warmer temperatures and only three articles (14 per cent) mentioned risk more explicitly.

In the articles about Arctic sea ice, 11 included general, unsourced statements mentioning the adverse impacts from Arctic sea ice melt, and ten included statements from named scientists/experts or scientific reports mentioning them. Only one article mentioned risk more explicitly.

All but one article (20) about Arctic sea ice mentioned the (implicit) risks posed to the changing local climate in the Arctic. Nine mentioned risks from changes in temperatures and eight mentioned effects on local ecosystems, including polar bears. The social dimension was not prominent. Two articles mentioned the effects of melting sea ice on local Inuit populations and only one mentioned the possibility of heightened political tensions in the region.

Three of the four articles where the opportunity frame was present in the coverage of the IPCC reports mentioned the possibility of warmer weather offering longer growing seasons. The ten articles with this frame in the Arctic sample included statements mentioning opportunities and/or possible positive impacts from sea ice melt. More than half of these articles pointed out the economic opportunities. Six articles mentioned improved chances of exploiting local resources such as minerals, oil, and gas, and six mentioned the opening of new shipping routes.

Looking at the entire Norwegian sample, explicit risk was found in only four articles (9 per cent). It was never salient but it was a dominant tone in two articles – one reporting from the WG-1 and the other from the WG-2. In these two articles, explicit risk provided a sustained framing alongside implicit risk and the verb 'to risk' ('å risikere') was used in both. The articles were published in the tabloids *VG* and *Dagbladet*. *VG* quoted a prominent Norwegian IPCC author and *Dagbladet* quoted international experts and the IPCC in general. Still, explicit risk was also found in several general, unsourced statements in both articles, perhaps as a way to create more drama.

Overall, explicit risk was more present in the tabloids' reporting than in *Aftenposten*. In total, seven articles (16 per cent) contained direct quotes about explicit risk by named scientists or scientific reports; five of them appeared in *VG* and *Dagbladet*. Considering that 58 per cent of our sample was taken from *Aftenposten*, this is a noticeable under-representation for that newspaper. The word risk occurred five times in the articles, always in tabloids, and metaphors were used in two articles, both of which appeared in *VG*.

There is no recognisable pattern distinguishing the reporting of the three papers. From the sample we can draw the following general observations:

- Journalists take cues from scientists when using the implicit risk frame.
- The implicit risk frame and uncertainty frame often appear in the same article.
- Norwegian media put a greater emphasis on the opportunities that follow from Arctic sea ice melt. This is in line with the domestic framing of a majority of the climate reporting found in other studies.
- Impacts on society received more attention than impacts on science when reporting the IPCC reports. When reporting the Arctic sea ice, changes to temperature and climate received overwhelmingly more attention than changes to society.
- Sceptical voices are largely absent.
- A significant amount of uncertainty in the reporting of the IPCC reports is due to mentions of uncertainties about climate science, future projections, range of temperature increases or uncertainty parameters.

United Kingdom
by James Painter

For several years, the UK has been a key player in the international politics of climate change. Under successive Labour governments and the Coalition government that replaced it in 2010, it has pushed for action at the UN, G-8 and EU levels, and has taken a lead in aiding developing countries to address climate issues. When the Climate Change Act was passed in 2008, the UK became the first country in the world to pass legislation to curb carbon emissions. It set itself the ambitious long-term domestic target of an 80 per cent reduction in all greenhouse gases by 2050. Under the Act, the government was also required to produce a Climate Change Risk Assessment, which it published in January 2012. It was the first such assessment of its type, and provided a wide-ranging and detailed overview of the impacts of climate change that pose the most urgent risks to the UK.

It is also a location where, like the USA, climate sceptics have sought to mobilise and influence public opinion against concerted action on climate change – not least with the formation in November 2009 of the Global Warming Policy Foundation (GWPF), a think tank which is active and vocal on the issue. The GWPF's chaiman, Nigel Lawson, is a former chancellor of the Exchequer (finance minister) under Margaret Thatcher. He and others on the right wing of the Conservative Party have had some success in mobilising grass-roots opposition to taking action on climate change, even though most of the leadership, including David Cameron, remained committed in principle.

Previous studies by the RISJ have found that the USA and UK have a significantly stronger presence of the different types of sceptical voices both in the media and wider society, compared to many other countries including France, Brazil, India, and China (although not necessarily other Anglo-Saxon countries like Australia and Canada) (Painter, 2011; Painter and Ashe 2012). A more recent study suggests that although the amount of coverage of climate change in the UK print media dropped off considerably in 2010–11 compared to 2009, there was only a small drop in the percentage of articles about climate change containing sceptical voices. It remained at about one in five articles (Painter and Gavin, 2013).

The constant presence of climate change sceptics in some of the right-leaning print media like the *Daily Express* and the *Sunday Telegraph* may be one of the reasons (there are many) why the UK often registers a significantly lower level of belief in either the seriousness or the anthropogenic nature of climate change, when compared to many other countries.

For example, in a 2012 AXA/Ipsos survey of 13,000 internet users in 13 countries, the UK came second only to the USA in having the highest number of respondents who believed that climate change was mostly due to natural factors:[11] 34 per cent in the UK believed this, and 42 per cent in the USA. This compared with percentages in Hong Kong (94 per cent), Indonesia (93 per cent), Mexico (92 per cent), and Germany (87 per cent) who believed that human activities were the main cause. A 2010 Gallup poll of more than 100 countries put it in 41st place, well behind such countries as Brazil, Japan, South Korea, and Canada. Fifty-seven per cent of those polled in the UK saw global warming as a 'very' or 'somewhat' serious threat.[12]

Many surveys suggest that the UK is like many other countries in that people have become less concerned about climate change in recent years and, most importantly for this study, more uncertain about the issue (Poortinga et al, 2011). Surveys carried out across Europe and the USA in 2009–10 found that scepticism and uncertainty had increased since the mid-2000s, perhaps as a result of the controversies surrounding leaked emails from scientists working at the University of East Anglia, errors made in the IPCC 2007 report on glacial melting forecasts in the Himalayas, and the unusually harsh European winter of 2009–10.

Specifically in the UK, a number of polls of public attitudes towards climate change documented an increase in the degree of perceived uncertainty about climate change over roughly the same period (BBC, 2010; Pew Research Center, 2009; Spence et al, 2010). One way of judging uncertainty is to test how much people believe the seriousness of climate change has been overplayed: the 2012 Shuckburgh study, which relied on interviews with around 1,000 respondents in the UK, found that around 44 per cent believed that it had indeed been exaggerated, a considerably higher figure than for the rest of Europe (27 per cent). It also found that the percentage of respondents who trusted independent scientists to tell the truth about climate change had fallen from 68 per cent in 2006 to 51 per cent in 2011 (Shuckburgh, Robison and Pidgeon, 2012: 12–14).

The UK media's coverage of climate change is very varied in terms of the amount of newsgathering resources different newspapers assign to it, the amount of space dedicated to it (including news and opinion) and the editorial line on it. The volume of coverage has peaked and troughed according to a wide variety of drivers such as the release of IPCC reports, the Copenhagen summit in 2009, and cold winters.[13] From 2004 to 2013, whereas the *Guardian* and the *Telegraph* provided regular coverage of the topic, the *Mail* registered one of the lowest amounts of coverage. A report published at the beginning of 2013 found that overall coverage of climate change in all English-language newspapers and agencies was slightly down compared to 2011, continuing a trend since 2009 (Fischer, 2013). Other studies suggest a slight increase, but more important for this study is that the *Guardian* and the *Telegraph* (along with the *New York Times*) were amongst the newspapers which covered climate change the most. Three of the most prolific 15 reporters (Fiona Harvey, Suzanne Goldenberg, and Damien Carrington) were from the *Guardian*. The *Daily Telegraph*'s environment correspondent, Louise Gray, also appeared high up the list (with 72 stories in 2012).

Indeed, since 2008 the *Guardian* has poured resources into covering climate change, partly in the belief that the scale of its resources should match the scale of the problem (Painter, 2011: 95). For some years, its print and online team consisted of six full-time environment correspondents, two editors, a dedicated picture editor, and two production journalists. Its website has a dedicated section on climate change with 28 outsider commentators regularly appearing alongside its various correspondents and editors. In October 2012, the paper's editor, Alan Rusbridger, said that its dedicated website was getting 2.4 million unique users a month, and registering 20 per cent annual growth. He said the *Guardian* was committed to covering climate change extensively because it was 'the most important issue of the age'.[14] Despite a reduction in the *Guardian* team in 2012, it still had by far the most resources of any UK media outlet dedicated to covering the topic.

At the time of writing, the *Daily Telegraph* has, unlike other newspapers like *The Times*, kept on its environment correspondent, backed up by its regular columnist and long-time environment correspondent Geoffrey Lean. The *Sunday Telegraph* regularly gives space to the prominent sceptic and author Christopher Booker.

For most of the period we examined, the *Mail* also had a dedicated environment editor who, like the *Telegraph*'s correspondent, tended to follow the mainstream scientific view on climate change, in contrast to many commentators in its opinion pages who do not. Indeed, the RISJ study on climate scepticism found that in the period 2009–10, the *Mail* and *Telegraph*s had a significantly higher percentage of uncontested sceptical opinion pieces or editorials compared to most other UK newspapers (with the exception of the *Sun* and the *Express*) (Painter, 2011: 102).

The *Mail*'s daily circulation is declining less rapidly than the *Telegraph* and *Guardian*. In December 2009 it was 2.1 million, compared to the *Guardian*'s 260,000 and the *Daily Telegraph*'s 630,000. By December 2012, it had fallen to 1.8 million compared to the *Guardian*'s 200,000 and the *Telegraph*'s 550,000. Perhaps most significantly, in 2012 the *Mail*'s online site climbed to being the most read 'news' English website in the world, overtaking the *New York Times*. In October that year, it had 50 million unique visitors compared to the *New York Times*' 49 million and the *Guardian*'s 39 million.[15] The Telegraph media group remains a successful one. It has the largest circulation in the UK market, the wealthiest readers, and the seventh most visited news website in the world. It made an operating profit of nearly £60 million in 2012.

While much of the Mail Online's content is celebrity gossip, it does carry news pieces on climate change and other harder news. Indeed, whereas the print version of the *Daily Mail* did not cover the record Arctic sea ice melt in the period August to September 2012, Mail Online did include three lengthy pieces. It was also significant that the *Telegraph's* website carried four news pieces on the record melt in this period, whereas the newspaper only ran one.[16] In contrast, the *Guardian* and *Observer* newspapers dedicated a very significant amount of space to the story, in part because one of its environment correspondents (John Vidal) was travelling with a Greenpeace team in the area.

The UK media's coverage of risk and uncertainty

The same five hypotheses as outlined in the previous chapter were examined in the case of the three UK newspapers. The results from the coding sheets can be seen in Table 6.5, and would suggest the following for the coverage of the three IPCC reports:

- The **uncertainty** frame was present in 26 of the 29 articles (90 per cent), although it was tempered by the 'increasing certainty' frame in nearly all (22) of those articles. It was not particularly salient (only three of the articles had it as a headline or dominant in the opening few lines). Nor was it the dominant tone (present in only four of the articles). However, it was to be found in several direct quotes (in 12 of the articles), and in the presence of duelling experts (six of the articles). It was of note that ten of the articles (around a third) had some mention of sceptical opinion.
- The **implicit risk** or disaster frame was particularly present. It was to be found in 24 of the 29 articles (84 per cent), and was particularly salient (in 14 of the articles, equivalent to nearly half). It was the dominant tone in 20 of the articles (69 per cent). Moreover, it was found in twice as many articles with direct quotes as the next most quoted frame. Adjectives such as 'catastrophic', 'severe' and 'devastating' were found in around a third of the articles.
- The **explicit risk** frame was much less present. Although it was to be found in nine of the articles, this was mostly related to the use of the word risk, rather than a sustained framing. Indeed, the word 'risk' was found 25 times throughout the articles. However, it was not particularly salient (four of the articles), nor was it the dominant tone (seven of the articles). Indeed, it was usually combined with the implicit risk

Table 6.5 UK Results

	Period 1 WG-1	Period 2 WG-2	Period 3 SREX	Total IPCC	%	Period 4 Arctic	%	Total	%
Number of articles	17	9	3	29		40		69	
Uncertainty	**15**	**8**	**3**	**26**	**90**	**30**	**75**	**56**	**81**
More certainty	15	7	0	22	76	20*	50	n/a	n/a
Duelling experts	4	2	1	6	21	n/a	n/a	n/a	n/a
Salience	1	2	0	3	10	8	20	11	16
Direct quotes	6	3	3	12	41	16	40	28	41
Word presence	3	0	4	7	n/a	3	n/a	10	n/a
Sceptics	6	2	2	10	34	4	10	14	20
Dominant tone	1	2	1	4	14	7	18	11	16
Implicit risk	**14**	**8**	**2**	**24**	**83**	**34**	**85**	**58**	**84**
Salience	10	3	1	14	48	18	45	32	46
Direct quotes	13	9	2	24	83	17	43	41	59
Adjectives	13	4	1	18	67	15	38	33	48
Metaphors	4	0	0	4	14	3	8	7	10
Dominant tone	12	6	2	20	69	24	60	44	64
Explicit risk	**3**	**5**	**1**	**9**	**33**	**5**	**13**	**14**	**20**
Salience	0	3	1	4	14	2	5	6	9
Direct quotes	1	4	1	6	21	1	3	7	10
Adjectives	2	1	0	3	10	0	0	3	4
Metaphors	0	0	0	0	0	0	0	0	0
Word presence	5	17	3	25	n/a	17	n/a	42	n/a
Dominant tone	2	4	1	7	24	2	5	9	13
Opportunity	**2**	**1**	**0**	**3**	**10**	**23**	**58**	**26**	**38**
Salience	0	0	0	0	0	5	13	5	7
Direct quotes	3	1	0	4	14	3	8	7	10
Word presence	1	0	0	1	n/a	7	n/a	8	n/a
Dominant tone	0	0	0	0	0	6	15	6	9
IPCC concepts	**8**	**3**	**2**	**13**	**45**	**n/a**			
Explanation	2	0	0	2	7	n/a			

* Year ice-free

frame when it was a dominant tone. Finally, the **opportunity** frame was neither very present, nor salient nor dominant.

- The **IPCC concepts** of likelihood and probability were present in 13 of the 29 articles, but in only two of them were the concepts explained.

Results from the coverage of the Arctic sea ice melt:

- The **uncertainty** frame was present in 30 of the 40 articles reviewed (75 per cent), but a considerable number of these (20 out of the 30) was due to the presence of the 'when will the Arctic be ice-free?' question. It was not particularly salient or a dominant tone (roughly one in five of the articles in each case). However, there were a significant number of articles with uncertainty quotes within them (16 of the 40).
- The **implicit risk** or disaster frame was particularly present. It was to be found in 34 of the 40 articles (85 per cent), and was particularly salient (in 18 of the articles, equivalent to nearly half). It was a dominant tone in 24 of the articles (60 per cent), and it was particularly found in the linguistic repertoire of adjectives such as 'devastating', 'catastrophic' and 'terrible'.
- The **explicit risk** frame was much less present. Only five of the 40 articles included it in some shape or form, and only two had it as a dominant tone. It was salient in only two of the articles.
- The **opportunity** frame was present in well over half of the articles (23). It was salient in five of them, and a dominant tone in six of them. The word 'opportunity' appeared seven times across the 40 articles, compared to just the once in the reporting of the three IPCC reports.
- Of the **specific possible impacts** mentioned in the articles, the effects on the local ecosystem, including polar bears, was the one mentioned most (in 18 of the articles), followed by possible colder winters in the northern hemisphere (13) and the effects on the ecosystem from mineral, gas, and oil exploration (11). The least mentioned was possible methane release (three). As regards possible opportunities, the improved chances of exploration were mentioned in 19 of the articles (almost half) and new shipping routes in 14.
- Seven of the articles included direct quotes from scientists expressing emotion at the state of the ice melt, which was surprise, shock or concern. This expression of personal feelings by scientists was regarded as both unusual and significant by some commentators (e.g. Porritt, 2012).

From a review of all the articles, the dominant picture is that of a very strong presence of both the uncertainty and implicit risk frames (81 per cent and 84 per cent respectively of the 69 articles examined). However, judged by other criteria, uncertainty was not a particularly significant frame (salient and dominant in just 16 per cent of articles). The presence of duelling experts was found in one in five of the articles about the IPCC reports, and presence of sceptics in one in five of all articles. Of the five articles in which sceptical voices represented more than 10 per cent of the content, four were to be found in the *Daily/Sunday Telegraph*, and three were opinion pieces (two authored by their regular sceptical columnist Christopher Booker, and another by the sceptical Australian scientist Bob Carter).

It is also worth pointing out that nearly 60 per cent of the articles included a quote from a scientist or scientific report with uncertainty contained within it in some shape or form. The actual use of the word 'uncertain' was not a particularly useful pointer. Indeed, the word was only present in six of the articles, even though the concept was present in 56 of them.

There was some variation between the three newspapers in the salience and dominance of uncertainty, but not in its presence. In all three newspapers, it was present in between 72 per cent and 84 per cent of the articles, with the *Guardian/Observer* having the highest presence. Measured by salience, the *Daily Mail/Mail on Sunday* had the highest percentage (25 per cent or two out of eight articles), compared to 14 per cent (four out of 21 articles) in the *Telegraph*s and 9 per cent in the *Guardian/Observer*. However, perhaps the most significant difference was found when uncertainty was judged by dominance: 24 per cent of articles in the *Telegraph*s had uncertainty as a dominant tone compared to 9 per cent in the *Guardian/Observer*. The difference was mostly explained by the presence of sceptical opinion articles in the former but not the latter.

As we saw in Chapter 4, other studies of the UK print media have shown that the disaster or alarmist frame is very common and the opportunity frame very uncommon. Our results would tend to support these conclusions for the reporting of the three IPCC reports. For example, the disaster framing was the dominant tone in nearly 70 per cent of the articles, salient in nearly half, and included a direct quote from experts in over 80 per cent. Strong adjectives were often used to emphasise the possible negative consequences. The words 'catastrophic' and 'devastating' were particular favourites. However, for the study of the Arctic sea ice melt coverage, the opportunity frame was much more present (58 per cent of articles), salient in 13 per cent, and a dominant tone in 15 per cent. This is to be expected given the obvious opportunities represented by oil, gas, and mineral exploitation and faster shipping routes. Two of the three articles from periods 1 to 3 where the opportunity frame was present mentioned the advantages of a move to cleaner technologies or a low-carbon economy.

The explicit risk frame was not common. It was found in 20 per cent of all the articles, salient in 9 per cent, and a dominant tone in 13 per cent. One would have expected a strong presence in the coverage of the SREX report for the reasons outlined in Chapter 5, but the sample size for the UK was small (only three articles). However, it was salient and dominant in one of these. It was surprising that the IPCC concept of likelihoods (when numbers were attached) was rarely present, which would have had the effect of boosting the presence of the explicit risk frame. Of the nine times explicit risk was a dominant tone in the articles, it was never the single dominant tone but was combined with the implicit risk frame on each occasion, and with the uncertainty and opportunity frame once each.

It is worth asking what were the main drivers or prompts for the presence of the explicit risk frame. Of the nine times it was a dominant tone, five were to be found in the reporting of the second IPCC report on impacts and the SREX report, which attracted phrases such as 'millions of people (or hundreds of species) at risk' from warming temperatures. Two were opinion pieces in period 1 which strongly used the language of risk. Perhaps the most interesting presence of the frame was an article in the *Guardian* in April 2012 about a report about the Arctic by Lloyd's of London, the world's largest insurance market, called 'Arctic opening: Opportunity and risk in the high north' (Kollewe and Macalister, 2012). Unsurprisingly, the article was full of the language of explicit risk, prompting the obvious point that journalists will often follow the tone, language, and framing of a press release or summary of a report.

This point can be endorsed by a brief examination of the reporting of the UK Climate Change Risk Assessment of January 2012 (CCRA), which also extensively used the language of risk and opportunity. In its press release, the CCRA mentioned the word 'risk' 19 times and 'opportunity' three times. A brief look at how the BBC website and the *Guardian* reported the release of the report suggests how they adopted this framing of climate change. The first line of the BBC report on 26 January 2012 stresses the risks and opportunities that climate change will bring in coming years, and also includes discussion of the uncertainties involved

(Shukman, 2012). Likewise, the *Guardian* report highlighted the risks and the opportunities, such as milder winters and drier summers, fewer cold-related deaths, better wheat crops, and a more attractive climate for tourists (Jowit, 2012). The word 'risk' was used five times in the one *Guardian* article – the same number of times the word appeared in the 17 articles we examined covering the release of the first IPCC report of 2007.

USA
by James Painter and Cassie Tickell-Painter

In a previous RISJ publication, *Poles Apart* (Painter, 2011: chapter 3), the argument was put forward that, in the context of climate change, the USA was 'exceptional' in various ways when compared to other industrialised countries. Four manifestations of this are a historical denial by some outlier scientists of the mainstream consensus on a variety of scientific issues, the financial and media muscle of think tanks, the presence of sceptical blogs, and the pervasive lobbying culture in the US body politic. But the USA also stands out from other countries in the correlation between public opinion about climate change and political sympathies, the high levels of sceptical opinion, and the large amount of space given to climate sceptics in the US media.

Climate change has long been a fault line in US politics, although over time there have been variations in the strength of this division. President Clinton agreed to ratify the Kyoto Protocol in 1997, but it was never taken to Congress during the two terms of President George W. Bush (2001–9) because he knew it would lose. However, towards the end of Bush's second term, there was considerable bipartisan agreement about the need to act on climate change. In 2008 the Democrat Nancy Pelosi and Republican Newt Gingrich recorded a film demonstrating bipartisan support for action, and in the election campaign the same year both the Republican presidential candidate, Senator John McCain, and the Democrat nominee, Congressman Barack Obama, agreed on the importance of tackling the issue.

However, the political landscape soon shifted again, and the attitudes of politicians towards climate change became more divided along party lines. In 2009 the Senate rejected Obama's proposal for a cap-and-trade scheme. On regaining a majority in Congress in 2010, the Republicans made disbanding the House Committee on Climate Change one of their first acts. Furthermore, Newt Gingrich publicly apologised in 2011 for the film with Nancy Pelosi calling it the 'single dumbest thing I've done in the last four years' (*Huffington Post*, 2011).

In the run-up to the 2012 election, climate change was remarkable for its absence as an issue. Obama failed to mention it in the 2011 State of the Union address. Furthermore, in all four formal debates during the campaign, none of the moderators asked a question on climate change, and no candidate mentioned it independently. Part of the reason for this lack of interest can of course be attributed to the greater priority of economic recovery and job creation. But climate change was in general not seen as a vote winner. Although Obama included US$90 billion for clean energy projects in his 2009 stimulus package, much of this money had been spent by 2011, and the promise of more money for climate projects at a time of high unemployment was not expected to play well with the electorate.

Obama's re-election in late 2012 marked another change. One week before the election, Hurricane Sandy hit the east coast and prompted a number of prominent politicians, from across the political spectrum, to highlight climate change. As mentioned in Chapter 1, the mayor of New York, Michael Bloomberg, famously announced his support for Obama – using the language of uncertainty and risk – in part because of Obama's policies towards climate change. On his re-election, Obama repeatedly brought up climate change, initially naming it as one of his three priorities for his second term. He said he would use his executive powers to tackle the issue rather than waiting for a deal in Congress, but at the time of writing he had failed to use those powers.

It is a particular characteristic of the USA that there is a strong correlation between public attitudes towards climate change and political affiliation. For example, a study by Michigan State University in 2010 found that whilst 70 per cent of Democrat voters saw human-caused warming as real, only 29 per cent of Republican voters did (McCright and Dunlap, 2011). A later study in 2011 found that whilst 78 per cent of respondents who vote Democrat believed that climate change was human-caused and action should be taken to combat it, only 53 per cent of Republican voters agreed. Furthermore, respondents who described themselves as belonging to the Tea Party movement were the least likely to agree that climate change was caused by humans: only 34 per cent of respondents agreed (Leiserowitz et al, 2011).

This division can in part be explained by different attitudes towards the role of government. Large-scale reductions in emissions (mitigation) can only really happen through federal government legislation so it is unsurprising that, in a country where many have a historical mistrust of big government, rejecting climate change is an appealing option. In many parts of the Republican right, the climate change issue has become a proxy for debates about big versus small government, or free market versus state intervention.

Opinion surveys have consistently shown some of the highest levels of scepticism about climate change in the USA. For example, a 2012 poll from the Yale Project on Climate Change Communication found that, for the first time since 2008, a majority of Americans agreed with the statement that climate change was both happening and caused by humans (54 per cent). This is in sharp contrast to most of the rest of the world, and indeed the USA often comes near, or at the bottom of, the list of countries believing in anthropogenic climate change.[17]

The US media coverage of climate change has both shaped and been shaped by the high level of polarisation in the debate. The abolition of The Fairness Doctrine in 1987, which required broadcasting companies to be 'honest, fair and balanced' in their presentation of controversial issues of public importance, paved the way for broadcasters with strong political affiliations such as Fox News and MSNBC. Fox News is well known for its constant espousal of a climate sceptic message. However, newspaper reporting of climate change does not have the same overt political affiliation as broadcast media. Unlike newspapers in other countries, US newspapers have largely maintained a tradition of keeping a firewall between comment and news.

Nonetheless, studies have shown that there is significantly more space given to sceptical voices, in all their manifestations, in the USA when compared to the rest of the world, with the exception of other Anglo-Saxon countries. The RISJ study *Poles Apart* concluded that the UK and the US print media quoted significantly more sceptical voices than the other four countries included (Brazil, China, France, and India). Together they represented more than 80 per cent of the times such voices were quoted. The study also showed that in these two countries there was a strong correspondence between the political leaning of a newspaper and its willingness to quote or use uncontested sceptical voices in opinion pieces and editorials.

The *New York Times* (*NYT*), *USA Today*, and the *Wall Street Journal* (*WSJ*) are the newspapers with the three highest circulations in the US. From April to September 2012 the *WSJ* had an average daily circulation of around 2.3 million (including digital editions), followed by *USA Today* with 1.7 million and the *NYT* with 1.6 million (and a Sunday circulation of 2.1 million).[18] Both the *NYT* and the *WSJ* are agenda-setting news outlets whose coverage strongly shapes the editorial decisions made by broadcast and cable networks. Although *USA Today* does not share the same reputation for excellence in journalism, it is the paper with the widest geographic circulation in the US.

The *New York Times* is known as a paper with a liberal stance, and is read by and directed at upper socio-economic groups. Historically it has had a dedicated and specialist team of science and environmental reporters. In 2009 they set up an environment desk incorporating the paper's own Green Blog, which had been created earlier as part of the energy 'cluster' at the Business desk. Its former environment correspondent, Andy Revkin, had already started the influential Dot Earth blog in 2007. However, in 2013 the environment desk was dismantled and the Green Blog closed.

Although the *Wall Street Journal* has a similar standing and socio-economic audience to the *NYT*, its political stance is right-leaning, and its coverage of climate change is quite different on its editorial and opinion pages (but not on its news pages). It has long given space to climate sceptics such as Rush Limbaugh, Fred Singer, and Bjørn Lomborg. The paper has been part of Rupert Murdoch's news empire since 2007, and it is in line with the climate sceptic tone of many of Murdoch's other papers, particularly in Australia. A 2012 study has shown the difference in tone in the *WSJ* and *NYT* as exemplified by their opinion pieces and editorials. In two three-month periods in 2007 and 2009–10, the *NYT* ran ten editorials relating to climate change, all of which were dismissive of climate sceptic arguments. In the same period the *WSJ* ran 12, only one of which seemed to be dismissive. The *NYT* also ran 14 opinion pieces, all of which contested the arguments of climate sceptics. The *WSJ* ran 17 opinion pieces, all but one of which was uncontested (Painter and Ashe, 2012).

In recent years, of the three papers in our study the *New York Times* has consistently given the most space to climate change coverage, whilst *USA Today* has provided the least. As in most countries the amount of coverage has shown sharp peaks and troughs. The peaks have been in 2007, coinciding with the publication of the IPCC reports, in late 2009 during the 'Climategate' affair and the Copenhagen summit, and in late 2012 in part driven by the extreme weather events.[19]

The US media's coverage of risk and uncertainty

Table 6.6 shows the results of the content analysis of the 55 articles in the sample. Articles included news stories in one of the paper's news sections, editorials, and opinion pieces. The results from coverage of the three IPCC reports are the following:

- The **uncertainty** frame was present in 25 of the 27 articles (93 per cent). Only eight of these mentioned increasing certainty. Whilst it was present in a high proportion of the articles, it was not particularly salient: only four of the articles had uncertainty in either the headline or first paragraph, or both. There were however a relatively high number of quotes from scientists conveying uncertainty – these were present in 15 articles or 56 per cent. Furthermore, sceptics were reasonably present as 11 of the articles included sceptical voices (41 per cent). Uncertainty was a dominant tone in just over a third of the articles (37 per cent).

- The **implicit risk** or disaster frame was present in a high proportion of the articles, though not quite as many as the uncertainty frame: 22 of the articles included reference to implicit risk (81 per cent). However, where it was present, the implicit risk frame was more salient than the uncertainty frame: 13 of the articles had implicit risk in either the headline or first paragraph (48 per cent). The proportion of articles that had implicit risk as a dominant tone was even higher: 19 of the 27 articles. Furthermore, a high proportion (70 per cent) of the articles included quotes relating to implicit risk. Strong adjectives were more common than for any other framing; examples were words such as 'biblical', 'powerful', and 'dire'.

- The **explicit risk** frame was not that common: it was found in ten of the articles (37 per cent). It was rarely salient, with only two articles including it. It was more commonly used as a dominant frame as nine articles included it, equivalent to a third of the sample articles. It was only once the dominant tone, and most likely to be used alongside implicit risk (eight times).

- The **opportunity** frame was not particularly present (22 per cent), was never salient, and was rarely the dominant tone (7 per cent).

- The **IPCC concepts** were used in 13 of the 27 articles, but were only explained in less than half of these: 22 per cent of all articles sampled.

Table 6.6 USA Results

	Period 1 WG-1	Period 2 WG-2	Period 3 SREX	Total IPCC	%	Period 4 Arctic	%	Total	%
Number of articles	16	8	3	27		28		55	
Uncertainty	**16**	**6**	**3**	**25**	**93**	**17**	**61**	**42**	**76**
More certainty	8	0	0	8	30	6*	21	n/a	n/a
Duelling experts	7	3	3	13	48	n/a	n/a	n/a	n/a
Salience	2	2	0	4	15	3	7	7	13
Direct quotes	11	2	2	15	56	6	21	21	38
Word presence	12	0	0	12	n/a	5	n/a	17	n/a
Sceptics	7	2	2	11	41	2	7	13	24
Dominant tone	7	2	1	10	37	5	18	15	27
Implicit risk	**12**	**7**	**3**	**22**	**81**	**22**	**79**	**44**	**80**
Salience	6	5	2	13	11	7	25	20	36
Direct quotes	10	6	3	19	70	9	32	28	51
Adjectives	9	4	2	15	56	10	36	25	45
Metaphors	0	1	0	1	4	3	11	4	7
Dominant tone	11	5	3	19	70	10	36	29	53
Explicit risk	**4**	**4**	**2**	**10**	**37**	**5**	**18**	**15**	**27**
Salience	0	1	1	2	7	3	11	5	9
Direct quotes	3	3	3	9	33	3	11	12	22
Adjectives	1	1	0	2	7	0	0	2	4
Metaphors	0	0	0	0	0	0	0	0	0
Word presence	9	16	2	27	n/a	12	n/a	39	n/a
Dominant tone	3	4	2	9	33	2	7	11	20
Opportunity	**2**	**4**	**0**	**6**	**22**	**18**	**64**	**24**	**44**
Salience	0	0	0	0	0	6	21	6	11
Direct quotes	2	4	0	6	22	2	7	8	15
Word presence	0	0	0	0	n/a	4	n/a	4	n/a
Dominant tone	1	1	0	2	7	11	39	13	24
IPCC concepts	**9**	**3**	**1**	**13**	**48**	**n/a**			
Explanation	6	0	0	6	22	n/a			

* Year ice-free

The results from the coverage of the Arctic sea ice melt are as follows:

- Although **uncertainty** was present in 61 per cent of the articles, it was rarely the dominant tone: only in five of the 28 articles or 18 per cent. Likewise, it was seldom salient: only three articles or 11 per cent of the sample. Furthermore, only 21 per cent of the articles had direct quotes by scientists about uncertainty. When uncertainty was mentioned, it was not principally in relation to the year the Arctic would be free of sea ice: only six of the 17 articles mentioning uncertainty referenced this. Most of the uncertainties related to the effects of sea ice melt, the most common being colder winters in the northern hemisphere.
- **Implicit risk** was present in a majority of the articles: 22 of the 28 sampled (79 per cent). However, it was not particularly salient (25 per cent) or often the dominant tone (36 per cent). Roughly a third of the articles included direct quotes about implicit risk, or used strong adjectives to describe the risks: 32 per cent and 36 per cent respectively.
- The **explicit risk** frame was far less present than either implicit risk or uncertainty: only five of the 28 articles included it (18 per cent). It was salient in three of these (11 per cent) and a dominant tone in two (7 per cent). It was never the sole dominant tone, and when it was strongly present, it was mixed with either opportunity or uncertainty.
- **Opportunity** was much more present in the articles about Arctic sea ice melt than in the IPCC articles: 18 of the 28 sampled included it (64 per cent). Eleven of these had opportunity as a dominant tone (39 per cent). It was salient in fewer articles (six).
- The most mentioned (implicit) risk was the possibility of colder winters in 13 of the articles (46 per cent), followed by political tensions in seven of the articles (25 per cent), and then effects on the local ecosystem in six articles (21 per cent). Methane release was only mentioned once.
- Exploiting resources was mentioned as a possible opportunity more frequently than new shipping routes, though both were quite frequent. The improved chances of mineral, oil, and gas exploration were mentioned in 17 of the articles (61 per cent), compared to the 12 times (43 per cent) the possibility of new shipping routes was discussed.
- Only two of the 28 articles included an emotional response from scientists. Both were from the *New York Times*. One expressed surprise and the other fear.

General results

The uncertainty frame had a very strong presence in all four periods (76 per cent), although it was more common in periods 1–3 (93 per cent) than period 4 (61 per cent). However, this reflected how frequently uncertainty was mentioned, rather than how salient it was in the article (just 13 per cent of the articles). It was a dominant tone in 15 of the 55 articles, equivalent to 27 per cent.

Although the uncertainty frame was not particularly salient or dominant, a considerable number of the articles, especially from the first three periods, contained sceptical voices, and quoted scientists expressing uncertainty in some way: 41 per cent of the articles from periods 1–3 included sceptical voices whilst 56 per cent included quotes on uncertainty. Fewer articles covering Arctic sea ice melt included either sceptics or uncertainty (7 per cent and 21 per cent respectively). The articles with more than 10 per cent given to sceptical voices were almost all from the *Wall Street Journal.* Of these, two were from period 1: one was written by a well-known climate sceptic, the British bio-geographer Philip Stott, and the other was an editorial.

The implicit risk frame was commonly found in all four periods (80 per cent of articles). It was the most salient of all four frames (36 per cent), had the largest number of articles with direct quotes from scientists (51 per cent), and the largest number of articles with strong adjectives (45 per cent). It was also the most dominant tone (53 per cent).

The opportunity frame was much less common, although it was found in 44 per cent of the articles. It was more frequently used in period 4, mentioned in 64 per cent of the articles compared to 22 per cent of the articles from periods 1–3. This was more than any other country studied. The six articles from periods 1–3 mentioned the advantages from longer growing seasons. Opportunity was a dominant tone in around two fifths of the articles from period 4, although less than one in ten for periods 1–3. This can be seen in the context of US claims to the Arctic. In the years from which the sample was taken (2010–12), the US and other Arctic nations like Russia, Denmark, and Canada were vying for rights for mineral extraction and shipping in the area. For example, the then Secretary of State Hillary Clinton became the highest ranking US official to take a seat on the Arctic Council in this period. This may help explain why opportunity was a more salient frame in the US than the other countries studied.

The explicit risk frame was not common in any of the periods. It was the single dominant tone once in all 55 articles. The other ten times it was a dominant tone it was combined with the three other frames and implicit risk the most. Explicit risk was often conveyed most through the use of the word 'risk'. For example, the article that included the explicit risk frame in the strongest way was published in the *New York Times* on 2 February 2007 and contained the word 'risk' four times (Rosenthal and Revkin, 2007). The article used the IPCC concepts of likelihood, and explained what the IPCC meant by them. It talked of the 'urgent need to limit looming and momentous risks' whilst also giving a numerical value (1 to 10) to the chance of even greater risks than those reported by the IPCC. An example of explicit risk came from a direct quote from the NOAA's Susan Solomon. However, implicit risk was also strongly present in the same article as the possible effects of climate change were covered, including hot extremes, heatwaves, heavy precipitation events, drought, and change in the pH of the oceans. Uncertainty was also present, mainly through explanations of where uncertainty remained and why it did so, including ranges in projections, and the possible effects of higher seas, sea ice melt or changes in rainfall patterns.

Explicit risk was a dominant tone in half of the articles from period 2, but only alongside one or more of the other frames. This was likely to be because the authors of the articles reflected the language of the WG-2 report. For example, one article from April 2007 in *USA Today* used the word 'risk' five times (Vergano and O'Driscoll, 2007). Furthermore, to explain how policy makers needed to respond to climate change, a comparison was made between designing buildings to withstand earthquake shocks and fire insurance policy. However, the framing of the article was still predominantly that of implicit risk, as it included a list of the likely effects of climate change in every continent. The article also contained uncertainty by posing the question in its headline 'Is Earth near its "tipping points"?', by giving a considerable amount of space to sceptics such as Bjørn Lomborg, by highlighting disagreement over the extent of the risk, and by including uncertainties around projections.

The *WSJ* gave significantly more coverage to Arctic sea ice melt than to the IPCC reports: ten articles from period 4 compared to a total of five from periods 1–3. The only article that included the 'explicit risk' frame was published in September 2011 (in period 4) and focused on the business opportunities (Gold, 2011). The headline highlighted both the opportunities and risks found in the article: 'Arctic riches lure explorers

– Exxon, Rosneft, Shell set to pour billions into potentially huge, risky prospects'. The risks included in the article were related to the effects of drilling in the Arctic, rather than directly to sea ice melt. The risk frame also came from the language of the companies. A Shell spokeswoman quoted in the piece said the 'risks are manageable' and the company plans to have oil-recovery vessels staged and ready to respond to any accidents. But again, overall, the article supports the general point that the explicit risk frame is most commonly found at the same time as other frames, rather than being dominant by itself. In this article, there was a mixture of opportunity and explicit risk, measured by presence, salience, and dominance.

It is interesting to note that the *New York Times* was more likely than the other newspapers in the sample to use the explicit risk framing. Explicit risk was a dominant tone in nearly half of the articles (six out of 13) it published on the three IPCC reports, compared to three out of nine for *USA Today* and none for the *WSJ*. Of the nine articles in periods 1–3 that included direct quotes from scientists using explicit risk language, seven were from the *NYT*, and of the 27 times the word 'risk' was used, 20 were found in the *NYT*. The *NYT* environment correspondent at the time, Andy Revkin, says he did find the risk language attractive:

> I was cutting back against frames such as climate crisis which was implying that everything was clearly established. To me it's not like that. We still don't know how much warming we are going to get from a doubling of CO2. Even the way the 2-degree threshold was derived, when you look carefully at it, it is used in a risk framework. But the way it is used is often in very certain terms that we need to cut emissions by 80 per cent by 2050 to avoid the end of the world – it's that kind of oversimplification that I found in my read of the science not justified. So I tried to move it to a better depiction of the problem of the ranges of what we don't know, using phrases like 'very likely' in the same way the IPPC did.[20]

7

Conclusions and Recommendations

As far back as 1990, the Conservative British Prime Minister, Margaret Thatcher, who was trained as a scientist and then firmly believed in the need to act on climate change, famously spoke of the dangers from global warming in a speech to the UN.[1] What is less known is that she used the language of risk and uncertainty. She thanked the UN for advancing the understanding of the risks of global warming, stressed that major uncertainties and doubts remained, but advocated that there was already a clear case for precautionary action at an international level. She had apparently borrowed the insurance concept from Ronald Reagan, who led negotiation of the 1987 Montreal Protocol to protect the ozone layer (Lochhead, 2013).

As we described in Chapter 1, more than 20 years later the mayor of New York, Michael Bloomberg, combined reference to an extreme weather event (Hurricane Sandy) with the risks and uncertainties surrounding climate change. In the UK, in February 2013 the Liberal Democrat secretary of state for energy and climate change, Ed Davey, used risk language to combat sceptics both inside and outside the governing coalition. He said that although there were uncertainties around some aspects of the science, 'those who deny climate change and demand a halt to emissions reduction and mitigation work, want us to take a huge gamble with the future of every human being on the planet, every future human being, our children and grandchildren, and every other living species. We will not take that risk.'[2]

There is considerable evidence that more political leaders and more science reports are now framing the climate change challenge in a broadly similar pattern, and this development was one of the main intellectual impulses for this study. One of its principal aims was to assess the extent to which the print media in different countries was beginning to reflect the increased usage of the language and concept of explicit risk. The implication of this study is that risk is not yet as embedded into climate change coverage as other strong narratives. This may be beginning to change as the coverage of the 2012 IPCC report on extreme weather events shows. Although this report was covered in far fewer articles than the 2007 IPCC reports, the explicit risk frame was present in half of them, and was relatively often a dominant tone.

However, in general very few articles in the periods we looked at included the explicit risk frame, or combined the ideas of explicit risk and uncertainty in the way that reports and some politicians are now doing. As can be seen from Chapters 5 and 6, there was a consistent picture across different countries, newspapers, and stories of the explicit risk frame being very subordinate to other frames.

The disaster/implicit risk frame, which stresses the negative impacts of climate change and in its extreme form is seen as an obstacle to public engagement with the issue, was overwhelmingly present, salient, and dominant across virtually all the newspapers we examined in the six different countries in the periods in question. Chapter 6 showed that there was evidence from the UK and Australia that the uncertainty frame, which highlights the various aspects which are not known about climate science and impacts, was more salient and dominant in the right-leaning than the left-leaning newspapers. But the overall picture across our sample was that the uncertainty frame, which is also often seen as a barrier to public understanding and engagement, was very present in all countries, although not so salient or dominant as the disaster frame.

This is not surprising. First, the story of climate change is essentially one of uncertain, but potentially very harmful, impacts. Secondly, the reports or events we chose to focus on had such elements in abundance. Thirdly, particularly when compared to positive stories of opportunity, the pull of the disaster frame is a strong one for journalists whatever culture they work in. As Meena Menon, the Mumbai Chief of Bureau for *The Hindu* neatly sums it up, 'I don't write alarmist stories but I can see how they would strike a chord in people. I would sit up and read such pieces. Also, the headline in the story of course is meant to attract readers – so you will find shocking statements. It is also written by someone else so there is bound to be a discrepancy between the headline and the text.'[3]

In many cases journalists, or more often headline writers, use hyperbolic disaster language when covering the news and science reports about climate change. The attraction of the 'bad news' story compared to the more positive story of opportunity is of course a strong shaper of the dominant narrative that readers or viewers receive. However, it should be stressed that, in much of their reporting, many journalists are also strongly guided by the main ways scientists and scientific reports present, or talk about, the science. As the analysis of the quotes found in the articles clearly shows, journalists are getting strong messages stressing uncertainties and negative impacts from scientists and scientific reports. With the SREX report, nearly half the articles contained quotes from the report or from scientists mentioning risk explicitly – this figure is much higher than other periods under examination and than the average figure. And as we saw in Chapter 6, although the UK's Climate Change Risk Assessment Report of January 2012 was not the main focus of the content analysis of this study, journalists (unsurprisingly) followed the main language of the report in stressing the risks and opportunities that climate change brings with it.

However, it is not always the case that journalists follow all the details of climate science reports. As outlined in Chapter 5, in the 177 articles we examined which covered the three IPCC reports, the concepts of likelihood and confidence levels were included in 79 of them (44 per cent), but an explanation of what they meant in only 27 (or 15 per cent). There may be various explanations for this, but an aversion to including (too many) numbers in each article is probably one of them.

This is important for several reasons. In the area of climate science, numbers and probabilities are likely to become more important in the coming years, not less. Even though some respected climate scientists argue that it is not possible (and unwise) to calculate accurate probabilities for possible outcomes,[4] it is very unlikely that the IPCC is going to stop using them in its reports. As climate models become more powerful and sophisticated, their potential to quantify uncertainties and generate probabilistic climate projections will be enhanced. In other words, uncertainty will be measured in more helpful ways as a basis for making decisions, even though it will not be eliminated. Giving ranges of probabilities and reliability or confidence levels will be an essential part of that. So it needs to be in the DNA of journalists to ask scientists or forecasters questions like 'what is the level of probability of such a weather event or impact happening, how confident are you in this prediction, and why are you confident?'

A related but different point is that considerable advances are also being made in the science of attribution or, in other words, working out through modelling the extent to which climate change altered the probabilities or odds of an extreme weather event having happened in the past. This could have huge implications for working out responsibility for damage claims, if they come to the courts. At the UNFCCC's December 2010 meeting in Cancun, loss and damage considerations were explicitly recognised, and a working group was set up that reported back at its next two meetings in Durban and Doha.

So a strong argument can be made that any journalist covering climate science will have to have a better understanding of risk, probabilities, and numbers. This will help considerably in providing a more constructive narrative about climate change than doom and gloom or uncertainty. In any event, as set out in Chapter 5, most journalists would benefit from having more familiarity with numbers. Away from the area of climate science, more data-focused reporting will become more common and necessary.

Specialist journalism is generally in decline in many Western countries, in part due to the problems facing the business model of print media. It is a worrying trend that most journalists are now generalists yet they have to cover highly specialised areas of risk in finance, health or the environment. In this context, it is hard not to see the reduction in specialist environment reporters as a worrying development that will impede better reporting of risk and uncertainty. In early 2013, the *New York Times* and *Washington Post* made cuts to the specialist coverage of climate change. The *NYT* shut down its Green Blog after previously closing its environment desk, while the *Post* switched its lead climate change reporter, Juliet Eilperin, away from environment reporting (although another reporter would take on the beat). CNN had cut its entire science unit several years previously (Donald, 2013). The BBC and some UK newspapers lost environment correspondents in 2012–13.

Some observers like *CJR*'s Curtis Brainard feel that the loss of specialist environment correspondents in the USA contributed to the uneven reporting of uncertainty around the wave of extreme weather events in 2012–13. 'Articles tended to bundle heatwaves, floods, and hurricanes into generalisations about extreme weather', he says.[5] 'Instead, they need to treat each type of event independently because each one comes with a different level of scientific evidence and agreement, from low to high, about the connection to climate change. Not to mention the fact that when there is agreement about a trend, it often applies only to a specific geographic region.'

Finally, it was not the purpose of this study to evaluate the possible impact on their readers and viewers of journalists' treatment of the stories in the four periods included in this sample. Our primary aim was to map what journalists and newspapers do, and understand some of the drivers. However, the media obviously play a huge role in providing science information to the general public and shaping the dominant messages they receive. Those who are interested in the effect on readers of certain messages about climate change will note the very strong presence of the disaster and uncertainty frames – the very frames which, as we saw in Chapters 2 and 3, at least in their more extreme forms are not usually seen as enhancing public understanding, engagement or behaviour change. Those who argue that including positive messages of opportunities arising from climate change is a more attractive option for public engagement will note the very low presence of this frame in the media coverage of the IPCC reports. The combination of using risk language and opportunity was virtually absent from our sample.

Recommendations

More familiarity and training for journalists about numbers and probabilities will improve coverage of climate risks.
There is already a wide variety of guides available for journalists and scientists on reporting or communicating risk and uncertainty around science in general, and climate science in particular.[6] But one gap in many of these guides is a lack of emphasis on the reporting of numbers and probabilities associated with the narrow (or statistical) sense of risk. As we have seen, many journalists have no background in science or mathematics, which can be – but not always – an obstacle to better reporting. Training sessions organised at the instigation of management or senior editors at media outlets is more likely to have an impact than the personal initiative of journalists.

Journalists cannot be expected to correct single-handedly the generally low level of public understanding of numbers, but as *CJR*'s Curtis Brainard says, 'As a profession, I wish journalists would go the extra steps, and not only report the news, but educate their readers and explain underlying concepts, such as confidence and uncertainty, in a scientific context. It's easier than ever to do this in digital media, where reporters have more space for text, graphics, and links.'[7]

There is more scope for inclusion in website articles of details and discussion about how uncertainty can be quantified and given a confidence level.

The idea that uncertainty can be quantified, and given a confidence level, and that it can be a form of information and not ignorance, is perhaps too much to expect journalists hampered by word lengths to take on board, but websites offer more opportunities. In this context, it is interesting to remember that experienced environment journalists welcomed the inclusion of probabilities and confidence levels in the IPCC reports.

The use of more info-graphics to illustrate the concepts of risk and other aspects of climate change needs to be (carefully) explored.

Although this has already been adopted by many media outlets, advances in technology should enhance their usage. The overwhelming majority of journalists interviewed for this study were enthusiastic supporters of more info-graphics, with the important caveats that they are only as good as the data behind them and that the data can be manipulated. More cross-fertilisation from the portrayal of risks in other areas of life such as health could also help.

The media could help the public to be more aware of probabilities by the greater use of probabilistic forecasting in public weather forecasting on television (in the UK and other countries where it is not being used).

This seems an ideal way of introducing the public to probability and risk around weather predictions which is clear, accessible and easily adaptable. Once the general public becomes more used to probabilistic forecasting, it should be easier for them to understand seasonal and long-range forecasts, and climate scenarios produced by modellers.

There is more scope for stories about risk and opportunity in specialist business magazines, whose readership understands these concepts as everyday concerns.

An October 2012 analysis by the Yale Climate Forum of business periodicals suggested that only the *Economist* consistently reported on climate change, disaster risk, and sustainability (Palmer, 2012). As the article argued, 80 per cent of the world's largest companies now published sustainability strategies to shareholders, reported on GHGs, or disclosed climate change risks. But 'while catastrophic risk and sustainability concerns associated with climate change now are increasingly reflected on corporate agendas, leading business magazines show little real appetite for substantive climate-related reporting'.

Scientists should stress early on in interviews with the media where there is broad consensus about climate science, and then later on where there are degrees of uncertainty. They should also try to explain that uncertainty does not usually mean ignorance.

More work is needed to explain to a non-specialist audience the uncertainties, and the probability and confidence levels, both in the IPCC and other science reports. Some scientists are already good at stressing what they know or are very sure about early on in interviews, before they discuss the areas of greater uncertainties. It is a thornier problem how to correct a common impression amongst the general public that uncertainty is the same as not knowing.

The IPCC 'blockbuster' Assessment Reports remain a very useful tool and peg for journalists around the world to give their readers and viewers a 'state of the art' report on where the latest knowledge of climate science is found. But the IPCC needs more resources to be able to communicate effectively around them and respond to media enquiries quickly.

Some scientists have argued that the IPCC blockbuster reports should be replaced by shorter, focused reports, with much smaller author teams, working entirely in public, addressing specific questions.[8] However, many environment journalists welcome these reports, and they are particularly helpful for journalists in developing countries (who form a majority of IPCC members).

However, it is inevitable that errors will creep in, and the media, encouraged by sceptics, will focus on these errors, so scientists and the IPCC need to be ready to accept errors, explain them, or put them in context. Many have argued that the enormous effort of producing the IPCC reports is not matched by resources for media responses and outreach work (see for example Ekwurzel, Frumhoff and McCarthy, 2012). For a long time, the IPCC only employed one full-time staff person to deal with the media, which at the time of writing had been increased to two (Painter, 2010: 79). Several of the journalists interviewed did note that the IPCC is improving its media work after a poor period of responding to media requests for clarity about aspects of their reports. But in the new digital age, the IPCC will come under more scrutiny and needs to be on the front foot in dealing with enquiries and challenges to its reputation.

Using the language of risk in the context of uncertainty can be a helpful way of presenting the problem to policy makers; but more research is needed about the effect on the general public of different types of risk language to test when it is effective, under what circumstances, with what groups, and with what metaphors.

Many of the journalists interviewed during the course of this study stressed the difficulties of communicating climate change in ways that help their audiences to understand the complexities and importance of it. It's a very knotty problem in part caused by the complexity of the science and the distance in time and space of the impacts and in part by the way everyone filters messages about climate change through their own value systems. There is no simple recipe or panacea to communicate it well. But risk has the obvious advantage of being a language common to other areas of life, and risk language is probably less of an obstacle to understanding and engagement than strong messages of uncertainty and future catastrophe. Risk can offer a more helpful and appropriate context in which to hold the debate about climate science and what to do about climate impacts. As Andy Revkin expresses it, 'communicating climate change is a devilish problem. And when you are facing a super-wicked situation, being less bad can be good'.[9]

Appendix 1

Coding sheet for coverage of IPCC reports

1. Name of researcher: ....
   Country: ....   Newspaper: ....

2. Period under research: (1, 2, 3) ....
   Search engine, search words, and search options used: ....
   ....

3. Total number of articles found when only relevant articles selected:
   ....

4. By article, in which part of the newspaper did the article appear?
   a. News reports and features: ....   b. Opinion pieces/columns:
   ....   c. Editorials: ....   d. Other: ....

5. Indicators of **uncertainty/certainty** frame:

   a. i. **General**: Did the article contain mentions of uncertainties about
          climate science, including future projections?
          Yes: ....   No: ....

      ii. Did the article contain such terms as 'increasing certainty',
          'increasing evidence that', 'almost certain', 'most certain to date',
          when referring to climate science, including future projections?
          Yes: ....   No: ....

   b. i. **Specific**: Did the article include uncertainty parameters or
          ranges in projections?
          Yes: ....   No: ....

      ii. Did the article mention the shortcomings of computer models?
      Yes: .... No: ....

      iii. Did the article mention the 'expansion of problem domain'
      (i.e. more research is needed)?
      Yes: .... No: ....

  c. **Tone**: Did the article have a presence of 'duelling experts' suggesting contestation about some aspect of the science?
  Yes: .... No: ....

6. Indicators of **risk** frame:

  a. Did the article include general, unsourced statements mentioning risks (broadly defined as possible adverse impacts or consequences) from warmer temperatures?
  Yes: .... No: ....

  b. Did the article include named sourced statements from scientists/ experts or scientific reports mentioning risks (possible adverse impacts or consequences) from warmer temperatures?
  Yes: .... No: ....

  c. Did the article include statements, either sourced or unsourced, mentioning risks more explicitly, either by the use of the word 'risk', 'precaution' or 'insurance', or where the odds, probabilities or chance of something adverse happening are given?
  Yes: .... No: ....

  d. Would you say that the article included more information on **scientific** impacts or consequences from warmer temperatures compared to **societal** impacts or consequences?
  Scientific: .... Societal: .... Not clear: ....

7. Indicators of **opportunity** frame:

  a. Did the article include statements, either sourced or unsourced, mentioning opportunities and/or possible positive impacts from warmer temperatures?
  Yes: .... No: ....

  b. Did the article include named sourced statements from scientists/ experts or scientific reports mentioning opportunities and/or

possible positive impacts from warmer temperatures?
Yes: .... No: ....

8. **Dominant** framing of article:

   a. **Salience of placing**: headline: is the headline predominantly one that contains: i. uncertainty: .... ii. risk: ....
   iii. opportunity: .... iv. none of these? ....
   If ii, is this risk 'implicit' or 'explicit'? Implicit: ...... Explicit: ......

   b. First paragraph (6 lines): is the first paragraph predominantly one that contains: i. uncertainty: .... ii. risk: ....
   iii. opportunity: .... iv. none of these? ....
   If ii, is this risk 'implicit' or 'explicit'? Implicit: ...... Explicit: ......

9   **Direct quotes**: are there direct quotes of named scientists or scientific reports which predominantly contain (a) uncertainty, (b) risk, or (c) opportunity?

   a. Uncertainty: Yes: .... No: ....
   If yes, how many different quotes? ....

   b. Risk: Implicit: Yes: .... No: ....
   If yes, how many different quotes? ....
   Explicit: Yes: .... No: ....
   If yes, how many different quotes? ....

   c. Opportunity: Yes: .... No: ....
   If yes, how many different quotes? ....

10. **Linguistic repertoire**:

   a. **Use of adjectives**: are there uses of strong adjectives ('enormous', 'large', 'substantial') which stress (i) uncertainty, (ii) risk, or (iii) opportunity?

   i. Uncertainty: Yes: .... No: ....
   If yes, give one example: ....

   ii. Risk: Implicit: Yes: .... No: ....
   If yes, give one example: ....
   Explicit: Yes: .... No: ....
   If yes, give one example: ....

  iii. Opportunity: Yes: ....   No: ....

   If yes, give one example: ....

 b. Is there a strong or dominant **metaphor** in the article which captures (i) uncertainty (ii) risk or (iii) opportunity?

  i. Uncertainty: Yes: ....   No: ....

   If yes, give one example: ....

  ii. Risk: Implicit: Yes: ....   No: ....

    If yes, give one example: ....

   Explicit: Yes: ....   No: ....

    If yes, give one example: ....

  iii. Opportunity: Yes: ....   No: ....

   If yes, give one example: ....

 c. **Relative frequency** of key words or phrases: doing a simple word count, how many times do the following words appear in the article? i. uncertain(ty): ....   ii. risk: ....
iii. opportunity: ....

11. **General tone/tenor?**

 a. In the article, is there a dominant tone or tenor to the article?
i. uncertainty: ....   ii. risk: ....
iii. opportunity: ....   iv. none of these: ....

  If (ii), is this risk 'implicit' or 'explicit'?
Implicit: ....   Explicit: ....

 b. Is there a key quote in the text which can be pulled out to give credibility to the choice of coding above? If so, what is it?

  ....

  ....

12. **Mainstream consensus versus sceptical voices:**

 a. Are sceptical voices included in the article which question some elements of mainstream climate science?
Yes: ....   No: ....

  How do they appear? i. sceptic as author of article: ....
ii. named sceptic(s): ....   iii. generic mention ('sceptics say that ...'): ....

b. If so, are they in the headline and/or first paragraph?
Yes: .... No: ....

c. Are they in direct quotes?
Yes: .... No: ....

d. Roughly what percentage of the article do their views represent?
0–10%: .... 10–50%: .... 50–90%: ....
90–100%: ....

e. If the answer to 12 (a) is yes, would you say that the article gives context about where the dominant scientific consensus lies?
Yes: .... No: ....

13. Did the article make use of the IPCC **concepts of likelihood** and confidence?

a. Yes: .... No: ....

b. If yes, did it mention the IPCC in connection with these concepts?
Yes: .... No: ....

c. If yes, did it explain what the IPCC means by these concepts?
Yes: .... No: ....

Coding sheet for coverage of Arctic sea ice melt

1–4 as in IPPC coding sheet.

5. Indicators of **uncertainty** frame:

a. **General**: did the article contain uncertainties about the future impact of Arctic sea ice melt?
Yes: .... No: ....

b. **Specific**:

i. Did the article include uncertainty about the year, or range of years the Arctic might be free of ice in the summer?
Yes: .... No: ....

ii. Did the article include uncertainty about the possible impact of the albedo effect on hastening the ice melt?
Yes: .... No: ....

iii. Did the article include other specific cases of uncertainty surrounding the impact of the future ice melt, other than (i) and (ii) above?

Yes: .... No: ....

If yes, please indicate what ....

....

6. Indicators of **risk** frame:

a. **General**:

i. Did the article include general, unsourced statements mentioning risks (broadly defined as possible adverse impacts or consequences) from Arctic sea ice melt?

Yes: .... No: ....

ii. Did the article include named sourced statements from scientists/experts or scientific reports mentioning risks (possible adverse impacts or consequences) from Arctic sea ice melt?

Yes: .... No: ....

iii. Did the article include statements, either sourced or unsourced, mentioning risks more explicitly, either by the use of the word 'risk', 'precaution', or 'insurance', or where the odds, probabilities or chance of something adverse happening are given?

Yes: .... No: ....

b. **Specific**: did the article mention the following (implicit) risks?

i. Changing local climate (in Arctic)

Yes: .... No: ....

ii. Possible changing climate in northern hemisphere (colder winters)

Yes: .... No: ....

iii. Possible methane release

Yes: .... No: ....

iv. Changes in (air and sea) temperatures in Arctic

Yes: .... No: ....

v. Effects on local ecosystem, including polar bears

Yes: .... No: ....

vi. Effects on local ecosystem from mineral/oil/gas exploration

Yes: .... No: ....

vii. Effect on lives of local (Inuit) people
Yes: .... No: ....

viii. Heightened political tensions in region
Yes: .... No: ....

7. Indicators of **opportunity** frame:

a. **General**: did the article include statements, either sourced or unsourced, mentioning opportunities and/or possible positive impacts from Arctic sea ice melt?
Yes: .... No: ....

b. **Specific**: did the article mention the following opportunities?

i. Improved chances of exploiting local resources (minerals, oil, gas)
Yes: .... No: ....

ii. Opening of new shipping routes
Yes: .... No: ....

8 and 9 (a) as in IPCC coding sheet.

9. b. Are there direct quotes from scientist(s) speaking of the Arctic ice melt where the scientist(s) talk of their emotional response?
Yes: .... No: ....

If yes, selecting the first quote in the article that appears, is the dominant emotion:
Fear: .... Worry: .... Sadness: ....
Shock: .... Surprise: .... None of these: ....

10, 11, and 12 as in IPCC coding sheet adapted to Arctic sea ice melt.

Appendix 2

Notes on methodology for content analysis

Word search: For periods 1–3, the search words were 'global warming' or 'climate change', at or near the start, and in some cases anywhere (in order to achieve a larger sample). All articles which were not principally about the IPCC report, or where the IPCC report did not provide the essential context for the article, were discarded. For the fourth period, the search words were 'Arctic' and 'ice' at or near the start, and in some cases anywhere. All articles which were not principally about the melting of the Arctic sea ice and global warming, or where the melting sea ice did not provide the essential context for the article, were discarded. All very short articles equal to or less than a couple of sentences long were also discarded, including trails. Letters, book reviews, film and TV reviews or corrections were also omitted. We did not examine the placing of articles within the newspaper, the length of the article, or the photos accompanying the articles.

For the **indicators of uncertainty** in section 5 of the coding sheet, we concentrated on a limited number of key indicators, including the following:

- Even though the word 'uncertain' did not often appear in the articles, words like 'may', 'suggest', 'likely', 'could' and 'possible' often did. Examples of the wide variety of phrases which indicated the presence of an uncertainty framing were 'warm spring *could* be due to …'; '*very difficult to predict* what will happen'; 'constantly humbled by *what we don't know*'; '*could be unfrozen*'; 'shipping rush *could* accelerate global warming'; 'multiple ways in which ecosystems *could* be affected'; 'North pole *could be* ice-free within next 10–20 years'; 'it's *unclear* what will happen'; '*potentially causing* big climatic changes'; 'scientists *don't yet know why*'; 'the final collapse *will probably be* complete in 2015–16'; and '*casting doubt on* predictions'.

- Uncertainty parameters or ranges in projections (e.g. for temperatures/ sea ice melt, year of ice-free Arctic), the mention of shortcomings of computer models, and the 'expansion of problem domain' (i.e. more research is needed) were also coded as indicators of the uncertainty frame. So, statements such as 'global temperatures could rise by 1.4– 3°C above levels for late last century' were coded as an uncertainty frame.
- Unsourced or sourced (from scientists or scientific reports) statements containing uncertainties about future projections and impacts were also coded. However, the uncertainty frame was often mitigated in period 1 by a 'growing certainty' sentiment, so this too was measured in the coding sheet. This was because the main angle of the IPCC WG-1 report was for many journalists that scientists were now more certain than before about the human causes of global warming.
- The presence of 'duelling experts' suggesting contestation about some aspect of the science was also coded. The quoting of sceptics either individually or generically was measured separately as a strong indicator of uncertainty.

In section 6 (d), examples of **scientific impact** were species loss, rainforest 'dieback', changing ecology of the Arctic, rising seas, more powerful hurricanes, more droughts and floods, disappearing coral reefs, or the presence of more thin ice; examples of **societal impact** would be mass migration, droughts or floods affecting people, food shortages, Arctic peoples being less able to hunt, more vegetables being able to be grown, and so on. Droughts and floods fitted both categories. If they were described in general terms, they were kept as examples of scientific impact, but if people/humans were mentioned, they were coded as societal impact.

In section 6, all 'expert' opinion was included. However, in section 9, only **quotes** from scientists or science bodies were counted. So this included the IPCC, and scientists working for universities or research bodies such as the Met Office, the EPA, and UNEP. Representatives of NGOs (for example Greenpeace or Oxfam), UN officials not associated with a research body (e.g. Ban Ki-moon), and politicians were not included. 'Direct quotes' meant a statement which was clearly assigned to a scientist or a science report, even though it may not have quotation marks around it. The IPCC reports were quoted several times in the same article, but these reports and others were only counted once for each category of uncertainty, implicit/explicit risk, and opportunity. Likewise, if the same scientist was quoted several times talking about implicit risks for example, this would only appear as once in the relevant column. However if five different reports or scientists were quoted on the same issue, this would appear as 5 in the same column. In the opinion pieces and editorials, we only included clear quotes from science reports or scientists. It was often the case that the authors of such pieces gave their own opinions which could be categorised as uncertainty or risk, but these were not included.

In section 10 (c), the **linguistic count** was normally carried out by a word search for the specific word. If these words came up in a context that was not linked to any aspect of climate change, then they were normally excluded.

In section 12, **sceptics** included direct mention of 1) trend sceptics (who deny the global warming trend); 2) attribution sceptics (who accept the trend, but either question the anthropogenic contribution saying it is overstated, negligent or non-existent compared to other factors like natural variation, or say it is not known with sufficient certainty what the main causes are); and 3) impact sceptics (who accept human causation, but claim impacts may be benign or beneficial, or that the models are not robust enough and/or question the need for strong regulatory policies or interventions).

In section 13, the relevant **IPCC terminology** is:

Confidence terminology	*Degree of confidence in being correct*
Very high confidence	At least 9 out of 10 chance
High confidence	About 8 out of 10 chance
Medium confidence	About 5 out of 10 chance
Low confidence	About 2 out of 10 chance
Very low confidence	Less than 1 out of 10 chance

Likelihood terminology	*Likelihood of the occurrence/outcome*
Virtually certain	> 99% probability
Extremely likely	> 95% probability
Very likely	> 90% probability
Likely	> 66% probability
More likely than not	> 50% probability

About as likely as not	33 to 66% probability
Unlikely	< 33% probability
Very unlikely	< 10% probability
Extremely unlikely	< 5% probability
Exceptionally unlikely	< 1% probability

Presence/salience/dominant tone

For the uncertainty results in Tables 5.3 and 6.1 to 6.6, the figures were taken from 5 (a) (i), 5 (b) (i), 5 (b) (ii) and 12 on the IPCC coding sheet, and 5 (a) and 5 (b) on the Arctic coding sheet; row 8 'more certainty' from 5 (a) (ii), and 'year ice-free' from 5 (b) on the Arctic coding sheet; row 9 from 5 (c); row 10 from 8 (a) and 8 (b); row 11 from 9; row 12 from 10; row 13 from 12; row 14 from 11 (with reference to 9 and 10).

For implicit risk, row 16 was taken from 6 (a) and 6 (b) (IPCC) and 6 (a) (Arctic); row 17 from 8 (a) and 8 (b); row 18 from 9; row 19 from 10 (a); row 20 from 10 (b); row 21 from 11 (with reference to 9 and 10).

For explicit risk, row 23 was taken from 6 (c) and 13 (IPCC) and 6 (a) (iii) (Arctic); row 24 from 8 (a) and 8 (b); row 25 from 9; row 26 from 10 (a); row 27 from 10 (b); row 28 from 10 (c), row 29 from 11 (with reference to 9 and 10).

For opportunity, row 31 was taken from 7; row 32 from 8 (a) and 8 (b); row 33 from (9); row 34 from 10 (c); row 35 from 11 (with reference to 9 and 10).

Rows 37 and 38 on the IPCC were taken from 13.

The 'word presence' count in rows 12, 28 and 34 is the total number of mentions of the word, not the total number of articles where the word appears.

Country variations

In four of the six countries (Australia, France, Norway, and the USA), we only included print versions of the articles. However, we had to use online versions in the Indian study because of the use of the newspapers' own search engines. We included in our UK sample seven online articles from August/September 2012 in period 4 from the *Telegraph* (four) and the *Mail* (three) because, if we had not, the sample from these two papers would have been very small for the record Arctic ice melt that year.

Coder reliability

All six coders were given the same coding sheets and the example of the UK coding results as a guide. The results from each country were checked by a second coder for consistency across all six countries. A representative sample of articles from the French and Norwegian newspapers was translated into English to check the coding.

Example of an article where three frames are dominant

An opinion piece in the *Sydney Morning Herald* on 6 April 2007, entitled 'Cool heads missing in the pressure cooker', had in its first line an implicit risk frame: 'The first thing that strikes you on reading the latest consensus report [...] is that it is like the plot of an Armageddon movie. "The climate of the twenty-first century is virtually certain to be warmer with changes in extreme events". A few lines later the article says 'the next thing that strikes you about the report is the high degree of uncertainty to which the authors readily confess'. This is an uncertainty frame. And a few lines later, the author quotes Lord Stern as saying 'What if that bet's wrong? You end up in a position that's extremely hard to extricate yourself [...] The basic economics of risk point very strongly to action.' This is an indicator of an explicit risk frame.

Notes

Executive Summary

1 Indicators of the **uncertainty frame** included ranges of projections, the presence of sceptical voices or duelling experts, and the inclusion of words like 'may', 'possible' or 'uncertain'. The **disaster or 'implicit risk' frame** included mention of possible adverse impacts such as sea level rises, more floods, water or food shortages, or population displacements, and, in the case of Arctic sea ice melt, negative effects on the ocean ecosystem and nations living on the Arctic rim, or heightened possibility of cold weather in the northern hemisphere. **Explicit risk** meant articles where the word 'risk' was used, where the odds, probabilities or chance of something adverse happening were given, or where everyday concepts or language relating to insurance, betting or the precautionary principle were included. A distinction was made between two types of **opportunity frame**: those accruing from doing something to reduce the risks from greenhouse gas emissions (the advantages of any move to a low-carbon economy), and those accruing from climate change (such as longer growing seasons in the northern hemisphere, or the prospects of new shipping routes and the possibility of mineral, gas, and oil exploration in the Arctic).

2 The countries studied were Australia, France, India, Norway, the UK, and the USA, and in each country three newspapers were chosen: normally, one right-leaning, one left-leaning, and either a tabloid or business newspaper. The four case studies examined were the first two IPCC Working Group reports of 2007, the IPCC report on weather extremes of 2012, and the recent melting of the Arctic sea ice.

3 The presence of a frame was measured by its inclusion anywhere within an article. Salience was measured by its presence within the headline or first few lines of an article. Dominant tones were assessed by considering various

criteria including weight of presence throughout an article, salience, prominent quotes, and the use of language such as metaphors and adjectives. An article may contain more than one frame.

Chapter 1   Introduction

1 www.wmo.int/pages/mediacentre/press_releases/pr_972_en.html.
2 For a discussion of 'framing', see Chapter 4.
3 Available via http://onlinelibrary.wiley.com/doi/10.1111/nyas.2010.1196.issue-1/issuetoc.
4 Email communication with author, March 2013.
5 See the video of Schneider called *Science and Distortion*, available via www.skepticalscience.com/science-and-distortion-stephen-schneider.html.
6 www.eci.ox.ac.uk/news/articles/LordDebenLecture.php.
7 Author interview, January 2013.
8 http://ipcc-wg2.gov/SREX/; www.defra.gov.uk/environment/climate/government/risk-assessment/#report.
9 http://climatechange.worldbank.org/content/climate-change-report-warns-dramatically-warmer-world-century.
10 Author interview, November 2012.
11 http://climatechangecommunication.org/sites/default/files/reports/Climate-Beliefs-September-2012.pdf.
12 Figure quoted by Professor Paul Hardaker at Royal Society meeting in London, 'Handling uncertainty in weather and climate prediction', 4 October 2012.
13 www.express.co.uk/news/uk/295296/Daily-fry-up-boosts-cancer-risk-by-20-per-cent.

Chapter 2   When Uncertainty is Certain

1 Author interview with Myles Allen, September 2012.
2 Stainforth, 2010, and author interview, November 2012.
3 Speaking at RISJ/GTC conference, 'Communicating Risk and Uncertainty', Oxford, 15 November 2012.
4 David Stainforth of the LSE is one of them. Author interview, November 2012.
5 This draws heavily on a chapter by Professor Stephan Harrison (2013).
6 www.skepticalscience.com/big-picture.html.

7 www.newscientist.com/special/climate-knowns-unknowns.

8 www.carbonbrief.org/profiles/what-we-know-and-what-we-dont.

9 http://royalsociety.org/uploadedFiles/Royal_Society_Content/policy/publications/2010/4294972962.pdf.

10 www.green-alliance.org.uk/uploadedFiles/Publications/reports/Climate ScienceBriefing_July11_sgl.pdf.

11 See David Spiegelhalter's lecture at Cambridge University, 3 May 2011.

12 Lecture given at Wolfson College, Oxford University, 10 May 2012.

13 Ekwurzel and Frumhoff are with the UCS, McCarthy is at Harvard University.

14 Chris Rapley speaking at RISJ/GTC conference, Oxford, November 2012.

15 Speaking at RISJ/GTC conference, Oxford, November 2012.

16 Email exchange with Richard Black, the BBC website's former environment correspondent.

17 For different ways of categorising sceptics see Painter, 2011, chapter 2.

18 Lawson was quoted or mentioned more than 50 times in the UK national press in a three-month period from November 2009 to February 2010, Plimer 12 times, and Inhofe five times.

19 'Infatuation with climate change is damaging recovery, says Lawson', *Daily Mail*, 11 June 2011. The *Australian* has quoted him expressing a similar sentiment – see 'Saving the planet will destroy the economy', *The Australian*, 25 June 2011.

20 Email correspondence with Catherine Happer, lead author of the report, *Climate Change and Energy Security*, UK Energy Research Centre.

21 Speaking at RISJ/GTC conference, Oxford, November 2012.

Chapter 3   The Language of Risk

1 As summarised by Lyn McGaurr and Libby Lester (2009).

2 Public lecture in London in February 2013, called 'Off the edge of history: the world in the 21st century', which can be heard via www2.lse.ac.uk/newsAndMedia/videoAndAudio/channels/publicLecturesAndEvents/player.aspx?id=1761.

3 See for example, Munich Re's October 2012 report on 'Severe weather in North America', and AXA, Risk education and research no 4, 'Climate risks', April 2012, available via www.axa.com.

4 www.qrisk.org.

5 www.defra.gov.uk/environment/climate/government/risk-assessment/#report.

6 http://ncadac.globalchange.gov.

7 http://climatecommission.gov.au/report/the-critical-decade.

8 Interview with author, December 2012.

9 Email exchange with Bob Ward, Grantham Research Institute, LSE.

10 Author interview, February 2013.

11 Author interview, November 2012.

12 Interviews with authors, January to March 2013.

Chapter 4   Reporting the Future

1 See for example Boykoff (2011: 189–218). Some individual country studies are included in chapter 7.

2 Speaking at RISJ/GTC conference, 'Communicating risk and uncertainty', Oxford, 15 November 2012.

3 Speaking at RISJ/GTC conference, Oxford, November 2012.

4 Spiegelhalter at RISJ/GTC conference, Oxford, November 2012.

5 See for example Spiegelhalter at http://understandinguncertainty.org/more-lessons-laquila.

6 Professor Steve Jones' report into the BBC can be found at www.bbc.co.uk/bbctrust/assets/files/pdf/our_work/science_impartiality/science_impartiality.pdf; Schneider (2009, chapter 7).

7 See discussion in Painter (2011, chapters 2 and 3).

8 Two examples are, from the USA, Nisbet (2009, p. 12), which outlines nine official and two suggested frames; and from the UK, Ereaut and Signit (2006), which divides articles into 12 linguistic repertoires.

9 Email exchange with Richard Black, a former environment correspondent at the BBC.

10 http://reutersinstitute.politics.ox.ac.uk/about/news/item/article/numbers-are-weapons-a-self-defenc.html.

11 Talk to RISJ fellows, as part of his Humanitas lectures, St Peter's College, Oxford, 5–9 November 2012.

12 www.guardian.co.uk/science/blog/2013/jan/11/science-on-journalism-curriculum.

13 www.nhs.uk/news/2012/12December/Pages/year-in-headlines-2012.aspx.

14 As observed by Professor David Spiegelhalter in his blog, 'Understanding uncertainty'.

15 www.rssenews.org.uk/2012/10/survey-of-mps-stats-literacy-hits-headlines.

16 Author interview, December 2012.

17 See for example University College London's Professor Nigel Harvey's (2012) letter to the *Guardian*.

18 www.metoffice.gov.uk/news/releases/archive/2011/weather-game.

19 Email communication with author, March 2013.

20 In the UK, one often proposed method is to give probabilistic information via the red button on the TV remote control.

21 See for example http://understandinguncertainty.org/visualising-uncertainty.

22 http://news.bbc.co.uk/1/hi/in_depth/629/629/6528979.stm.

23 See www.abc.net.au/news/2013-01-14/how-will-rising-seas-impact-australia/4460688.

24 Email interview, February 2013.

25 http://dotearth.blogs.nytimes.com/2009/02/23/warming-embers-burning-brighter.

26 http://globalchange.mit.edu/focus-areas/uncertainty/gamble.

27 http://blog.ucsusa.org/extreme-weather-and-climate-change; and http://ipcc-wg2.gov/SREX/images/uploads/SREX_SPM_BOX2.jpg.

28 Email communication with author, February 2013.

Chapter 5   Uncertainty and Risk in the International Print Media

1 See Max Boykoff's graphs at http://sciencepolicy.colorado.edu/media_coverage.

2 www.wmo.int/pages/mediacentre/press_releases/documents/WMO_1108_EN_web.pdf.

3 For details, see the RISJ Digital News Report 2012 (Newman, 2012).

4 Four million of this figure corresponds to *The Times of India*, and 1.5 to *The Hindu*. The total figure for the US newspapers was about 4 million, for the UK 2.5 million, and for Australia, France and Norway around 1 million each.

5 Author interview, March 2013.

6 2 February 2007 for WG-1 issued in Paris, 6 April 2007 for WG-2 issued in Brussels, and 28 March 2012 for the SREX report issued in Geneva.

7 WG-1: www.ipcc.ch/pdf/press-releases/pr-02feburary2007.pdf; SREX: http://ipcc.ch/news_and_events/docs/srex/srex_press_release.pdf.

8 WG-1: www.unep.org/documents.multilingual/default.asp; WG-2: www.wmo.int/pages/mediacentre/press_releases/pr_776_en.html, also available via UNEP site.

Chapter 6   Country Studies

1 For related discussion of the *Australian*'s use of 'cool heads' framing, see McGaurr and Lester (2009: 183).

2 The Eurobarometer surveys can be found via the website: http://ec.europa.eu.

3 www.ipsos.fr/sites/default/files/attachments/international-monitor-on-climate-risks-ipsos-axa.pdf.

4 The figures in 2011 were *Le Monde*, 325,295; *Le Figaro*: 334,406 in 2011; *Le Parisien*: 290,348 and *Aujourd'hui en France*: 173,984. Source: www.ojd.com/adherent/3147.

5 Author interview, March 2013.

6 Neither study categorises the sample based on their geography, economic background or whether they are from urban or rural India. For a country as large and diverse as India, with varied sources of information and access to them, this is an important consideration.

7 Author interview, March 2013.

8 www.thehindu.com/navigation/?type=static&page=aboutus.

9 Charts are available at http://sciencepolicy.colorado.edu/media_coverage/india/index.html.

10 Circulation figures for all newspapers are taken from MEDIEBED-RIFTENE (2012).

11 www.axa.com/lib/axa/uploads/cahiersaxa/Survey-AXA-Ipsos_climate-risks.pdf.

12 www.gallup.com/poll/147203/Fewer-Americans-Europeans-View-Global-Warming-Threat.aspx#2.

13 See the charts prepared by the CSTPR at the University of Colorado, Boulder, available via http://sciencepolicy.colorado.edu/media_coverage.

14 Talk at Frontline Club, London, 2 October 2012.

15 www.comscoredatamine.com/2012/12/most-read-online-newspapers-in-the-world-mail-online-new-york-times-and-the-guardian.

16 One of the *Guardian*'s columnists on the environment, George Monbiot, complained that most of the UK newspapers ignored the record levels of ice melt (Monbiot, 2012).

17 See for example the AXA/IPSOS survey of 2012, available at www.axa.com/lib/axa/uploads/cahiersaxa/Survey-AXA-Ipsos_climate-risks.pdf.

18 See the Audit Bureau of Circulations at http://abcm.org.my.

19 See the charts prepared by the CSTPR at the University of Colorado, Boulder, available via http://sciencepolicy.colorado.edu/media_coverage.

20 Author interview, March 2013.

Chapter 7   Conclusions and Recommendations

1 Speech at the Second World Climate Conference in Geneva on 6 November, 1990, available at www.margaretthatcher.org/document/108237.
2 www.gov.uk/government/speeches/edward-davey-speech-to-the-avoid-symposium-at-the-royal-society.
3 Interview with India researcher, March 2013.
4 See for example Professor Carl Wunsch at the MIT, quoted in Revkin (2007).
5 Author interview, March 2013.
6 See Bibliography.
7 Author interview, March 2013.
8 Professor Myles Allen cited in Dickson (2012). See wider discussion at www.bbc.co.uk/blogs/thereporters/richardblack/2010/02/rejigging_the_climate_panel.html.
9 Author interview, March 2013.

Bibliography

Sources cited

Allen, Myles, 'Al Gore is doing a disservice to science by overplaying the link between climate change and weather', Guardian online, 7 October 2011.

Aram, Arul, 'Indian media coverage of climate change', *Current Science* 100 (10), 25 May 2011.

Asdal, Kristin, 'Klimasakens forvaltning', *Dagens Næringsliv*, 20 April 2012.

Ashe, Teresa, 'How the media report risk and uncertainty around science – a review of the literature', *RISJ Briefing Paper*, forthcoming 2013.

Auletta, Ken, 'Citizens Jain: Why India's newspaper industry is thriving', *The New Yorker*, 8 October 2012.

Australian, 'Signals are getting warmer on climate', *Australian*, 9 April 2007, p. 9.

Aykut, Stefan Cihan, et al, 'Climate change controversies in French mass media 1990–2010', *Journalism Studies* 13 (2) (2012), pp. 157–74.

Bacon, Wendy, *A Sceptical Climate: Media Coverage of Climate Change in Australia 2011: Part 1 – Climate Change Policy*, Australian Centre for Independent Journalism, 2011.

Baume, Patrick, 'Top 10 stories of 2009', *Media Monitors*, 12 December 2009.

BBC, *BBC Climate Change Poll – February 2010*, 2010, available at http://news.bbc.co.uk/nol/shared/bsp/hi/pdfs/05_02_10climatechange.pdf.

BBC World Service, 'All countries need to take major steps on climate change: Global poll', 25 September 2007.

Bernstein, Peter L., *Against the Gods – The Remarkable Story of Risk*, New York: John Wiley, 1996.

Bevege, Alison, 'Claims climate "myth" may wreck economy', *Northern Territory News*, 11 June 2008.

Billet, Simon, 'Dividing climate change: global warming in the Indian mass media', *Climatic Change* 99 (1–2) (2009), pp. 1–16.

Boenker, Karyn, 'Communicating global climate change: Framing patterns in the US 24-hour news cycle, 2007–2009', unpublished thesis, University of Washington, 2012.

Bolt, Andrew, 'Andrew Bolt: Profile', *Herald Sun*, 2013.

Borenstein, Seth, 'America set an off-the-charts heat record in 2012', *Associated Press*, 8 January 2013.

Boyce, Tammy, *Health, Risk and News: The MMR Vaccine and the Media*, New York: Peter Lang, 2007.

Boyce, Tammy and Justin Lewis (eds), *Climate Change and the Media*, London: Peter Lang, 2009.

Boykoff, Max, 'The cultural politics of climate change discourse in UK tabloids', *Political Geography* 27 (5) (2008), pp. 549–69.

Boykoff, Max, 'Indian media representations of climate change in a threatened journalistic ecosystem', *Climatic Change* 99 (2010), pp 17–25.

Boykoff, Max, *Who Speaks for the Climate? Making Sense of Media Reporting on Climate Change*, Cambridge: Cambridge University Press, 2011.

Boykoff, Max and Jules Boykoff, 'Climate change and journalistic norms: A case-study of U.S. mass-media coverage', *Geoforum* 38 (6) (2007), pp. 1190–204.

Briggs, Chad Michael, 'Climate security, risk assessment and military planning', *International Affairs* 88 (5) (2012), pp. 1049–64.

Brossard, Dominique, James Shanahan and Katherine McComas, 'Are issue-cycles culturally constructed? A comparison of French and American coverage of global climate change', *Mass Communication and Society* 7 (3) (2004), pp. 359–77.

Brulle, Robert J., et al, 'Shifting public opinion on climate change: An empirical assessment of factors influencing concern over climate change in the U.S., 2002–2010', *Climatic Change* 114 (2) (2012), pp. 169–88.

Budescu, David, Stephen Broomell and Han-Hui Por, 'Improving communication of uncertainty in the reports of the IPCC', *Psychological Science* 20 (2009), pp. 299–308.

Bureau of Meteorology, 'Bureau of Meteorology confirms it's been the hottest summer on record', media release, Bureau of Meteorology, 1 March 2013.

Byles, Dan, 'UK energy and climate change policy – A few facts', *Huffington Post*, 4 February 2012.

Carrington, Damian, 'Climate change doubled likelihood of devastating UK floods of 2000', *Guardian*, 16 February 2011.

Chalmers, John, 'Special: Top 10 issues in the Australian media 2012', *Media Monitors*, 20 December 2012.

Clark, Pilita, 'World Bank chief makes climate change plea', *Financial Times*, 18 November 2012.

Clark, Pilita, 'Frozen frontiers', *Financial Times*, 7 February 2013.

Climate Commission, *The Critical Decade: Climate Science, Risks and Responses*, Climate Commission, 2011.

Climate Commission, *Climate Commission: The Year in Review 2012*, Climate Commission, 2012.

Cohen, Tamara, 'New Met Office forecast system likely to mean 80% chance of confusion', *Daily Mail*, 10 November 2011.

Corbett, Julia and Jessica Durfee, 'Testing public (un)certainty of science', *Science Communication* 26 (2) (2004), pp. 129–51.

CRED, *The Psychology of Climate Change Communication: A Guide for Scientists, Journalists, Educators, Political Aides, and the Interested Public*, New York, 2009.

Cubby, Ben, 'More wild weather on the way, UN climate panel says', *Sydney Morning Herald*, 19 November 2011.

Department of Climate Change and Energy Efficiency, *Australian National Greenhouse Accounts: Quarterly update of Australia's national greenhouse gas inventory: September quarter 2012*, Department of Climate Change and Energy Efficiency, 2012.

Dickson, David, 'IPPC should drop its big reports, meeting hears', *Scidev.net*, 22 November 2012.

Ding, Ding, et al, 'Support for climate policy and societal action are linked to perceptions about scientific agreement', *Letter to Nature*, December 2011.

Donald, Ros, 'What's the future of climate coverage?', *Carbon Brief*, 23 January 2013.

Doulton, Hugh and Katrina Brown, 'Ten years to prevent catastrophe?', *Global Environmental Change*, 19 (2009), pp. 191–202.

Duarte, Katherine, 'En ubehagelig sannhet om norsk klimadekning. Hvilke stemmer og holdninger blir representert i norsk pressedekning av klimaendringer', Master's thesis, University of Bergen, 2010.

Duarte, Katherine, 'Skeptical voices in Norwegian climate coverage', Poster presented at Planet Under Pressure Conference, 26–29 March 2012, London, UK.

Duarte, Katherine and Dimitry Yagodin, 'Scientific leaks: Uncertainties and skepticism in climate journalism', in E. Eide and R. Kunelius (eds), *Media Meets Climate: How Journalism Across the World Covers Climate Change Summits*, Gothenburg: Nordicom, 2012.

Economist, 'The vanishing north', *Economist*, 16–22 June 2012.

Economist, 'Climate science: A sensitive matter', *Economist*, 30 March 2013.

Eide, Elisabeth, 'Climate reporting – a global snapshot', Carbonundrums conference, Oslo, 2011.

Eide, Elisabeth and Andreas Ytterstad, 'The tainted hero: Frames of domestication in Norwegian press representation of the Bali climate summit', *The International Journal of Press/Politics* 16 (2011) pp. 50–74.

Ekwurzel, Brenda, Peter Frumhoff and James McCarthy, 'Climate uncertainties and their discontents: increasing the impact of assessments on public understanding of climate risks and choices', *Climatic Change* 108 (2012), pp. 791–802.

Entman, Robert M., 'Framing: Toward clarification of a fractured paradigm', *Journal of Communication* 43 (4) (1993), pp. 51–8.

Ereaut, Gill and Nat Signit, *Warm Words*, IPPR: London, 2006.

Fischer, Douglas, 'Climate coverage, dominated by weird weather, falls further in 2012', *Daily Climate*, 2 January 2013.

Flannery, Tim, Roger Beale and Gerry Hueston, *The Critical Decade: International Action on Climate Change*, The Climate Commission Secretariat, 2012.

Glasgow Media Group, *Climate Change and Energy Security: Assessing the Impact of Information and its Delivery on Attitudes and Behaviour*, UK Energy Research Centre, London, 2012.

Gold, Russell, 'Arctic riches lure explorers – Exxon, Rosneft, Shell set to pour billions into potentially huge risky prospects', *Wall Street Journal*, 1 September 2011.

Government of India, Ministry of Environment and Forests, letter from Joint Secretary, Ministry of Environment and Forests, to Executive Secretary, United Nations Framework Convention on Climate Change, 2010.

Government of Norway, 'State budget for 2013', 2013, available at www.statsbudsjettet.no/Statsbudsjettet-2013/Statsbudsjettet-fra-A-til-A/Kyoto-forpliktelsen/?query=kyoto.

Gray, Louise, 'Millions at risk from climate-change heat waves, UN warns', *Daily Telegraph*, 19 November 2011.

Gyngell, Allan, *The Lowy Institute Poll 2007: Australia and the World: Public Opinion and Foreign Policy*, The Lowy Institute for International Policy, 2007.

Hanson, Fergus, *The Lowy Institute Poll 2012: Public Opinion and Foreign Policy*, The Lowy Institute for International Policy, 2012.

Harrabin, Roger, et al, 'Health in the news', *King's Fund*, London, September 2003.

Harrison, Stephan, 'Climate change, uncertainty and risk', in Sally Weintrobe (ed.) *Engaging with Climate Change*, Hove: Routledge, 2013.

Hartcher, Peter, 'Cool heads missing in the pressure cooker', *Sydney Morning Herald*, 6 April 2007.

Harvey, Fiona, 'Extreme weather will strike as climate change takes hold, IPCC warns', *Guardian*, 19 November 2011.

Harvey, Professor Nigel, letter to the *Guardian*, 29 October 2012, available at www.guardian.co.uk/world/2012/oct/29/storms-rage-laquila-sentences.

Hassol, Susan Joy, 'Improving how scientists communicate about climate change', *EOS*, 89 (11) (2008), pp. 106–7.

Hellevik, Ottar, 'Miljøholdninger fra 1985 til i dag, sett i lys av den allmenne verdiutviklingen', presentation at the Klimaforum, Oslo, Norway, October 2010.

Henderson-Sellers, Ana, in Paul Barclay, '2010 George Munster Award forum', *Big ideas*, ABC Radio National, 4 November 2010.

The Hindu, 'Climate change will hit the poor hard, says report', *The Hindu*, 7 April 2007.

Hope, Chris, 'The social cost of CO2 from the PAGE09 model', *Judge Business School Working Paper 5* (2011), pp. 12–14.

Howard-Williams, Rowan, 'Ideological construction of climate change in Australian and New Zealand newspapers', in Tammy Boyce and Justin Lewis (eds), *Climate Change and the Media*, London: Peter Lang, 2009.

Huffington Post, 'Newt Gingrich: Nancy Pelosi climate change ad is "Dumbest thing I've done in the last 4 years"', *Huffington Post*, 27 December 2011.

Hulme, Mike, *Why We Disagree about Climate Change: Understanding Controversy, Inaction and Opportunity*, Cambridge: Cambridge University Press, 2009a.

Hulme, Mike, 'Mediated messages about climate change', in Tammy Boyce and Justin Lewis (eds), *Climate Change and the Media*, London: Peter Lang, 2009b.

Inhofe, James, 'First four pillars', 8 April 2005, Senate Floor Statement, available at http://inhofe.senate.gov.

IPCC, 'Working Group I contribution to the IPCC Fifth Assessment Report, Climate Change 2013: The physical science basis: Coordinating lead authors, lead authors and review editors', IPCC, n. d.

Jaitely, Arun, 'Cancun climate change negotiations: Alteration of India's stand and its impact', 2010, available at www.arunjaitley.com/en/my-opinion-inside.php?id=49&mode=Read&icatId=29.

Jarrett, Wendy, Associate Director, Media Relations, NICE, speaking at RISJ/GTC conference, Oxford, November 2012.

Jogesh, Anu, 'A change in climate? Trends in climate change reportage in the Indian print media', in Navroz K. Dubash (ed.), *Handbook of Climate Change and India: Development, Politics and Governance*, New Delhi: Oxford University Press, 2011, pp. 266–86.

Jones, Cheryl, 'American don sceptical about analysis of newspaper bias on carbon policy', *Australian*, 30 April 2012, p.28.

Jowit, Juliette, 'Flooding rated as worst climate change threat facing UK', *Guardian*, 26 January 2012.

Karoly, David, Matthew England and Will Steffen, 'Off the charts: Extreme Australian summer heat', Climate Commission, 2013.

Khan, Shabana and Ilan Kelman, 'Progressive climate change and disasters: communicating uncertainty', *Natural Hazards* 61 (2) (2012), pp. 873–7.

Kitzinger, Jenny, 'Researching risk and the media', *Health, Risk and Society* 1 (1) (1999), pp. 55–69.

Kollewe, Julia and Terry Macalister, 'Arctic oil rush will ruin ecosystem, warns Lloyd's of London', *Guardian*, 12 April 2013.

LaFountain, Courtney, 'Health Risk Reporting', *Social Science and Public Policy* 42 (1) (2004), pp. 49–56.

Le Monde, 'Menaces sur les grands fleuves', *Le Monde*, 6 April 2007.

Le Treut, Hervé, 'Que fait la France?', *Le Monde*, 30 January 2007.

Leiserowitz, Anthony, Edward Maibach, Connie Roser-Renouf, and Jay D. Hmielowski, 'Politics and global warming: Democrats, Republicans, Independents, and the Tea Party.' Yale University and George Mason University. New Haven, CT: Yale Project on Climate Change Communication, 2011, available at http://environment.yale.edu/climate/files/PoliticsGlobalWarming 2011.pdf.

Leiserowitz, Anthony and Jagadish Thaker, 'Climate change in the Indian mind', Yale Project on Climate Change Communication, Yale University, 2012.

Lertzman, Renee, speaking at UCL conference, 'Psycho-social dimensions of climate change', 20 November 2012.

Leys, Nick, 'Newspaper rejects carbon tax criticism', *Australian*, 2 December 2011, p. 2.

Lindsay Griffin, Erik and Lise Campbell Lehne, *Den Store Norske Klima – og Miljøundersøkelsen*, 2010.

Lloyd, Graham, 'Physicist mocks climate "scam"', *The Australian*, 14 October 2010.

Lochhead, Carolyn, 'On climate, GOP turns from concern to denial', *Houston Chronicle*, 26 April 2013.

Mabey, Nick and Katherine Silverthorne, 'Degrees of risk: Defining a risk management framework for climate security', E3G, London, February 2011.

MacAskill, Ewen and Suzanne Goldenberg, 'New York's Bloomberg endorses Obama to lead on climate change', *Guardian*, 2 November 2012.

Manne, Robert, 'Bad news: Murdoch's Australia and the reshaping of the nation', *Quarterly Essay* 43 (1 September 2011), p. 43.

Marshall, Michael, 'Arctic thaw may be first in cascade of tipping points', *New Scientist*, 27 February 2013.

Marshall, Michael and Fred Pearce, 'What leaked IPCC reports really says on climate change', *New Scientist*, 17 December 2012.

Mastrandrea, Michael D., et al, 'Guidance note for lead authors of the IPCC Fifth Assessment Report on Consistent Treatment of Uncertainties', 2012, available at https://docs.google.com/file/d/0B1gFp6Ioo3akNnNCa VpfR1dKTGM/edit?pli=1.

McCright, Aaron M. and Riley E. Dunlap, 'The politicization of climate change and polarization in the American public's views of global warming, 2001–2010', *The Sociological Quarterly* 52 (2) (2011), pp. 155–94.

McGaurr, Lyn and Libby Lester, 'Complementary problems, competing risks', in Tammy Boyce and Justin Lewis (eds), *Climate Change and the Media*, London: Peter Lang, 2009.

MEDIEBEDRIFTENE, 'Når flere lesere med redaksjonelt innhold', 2012, available at www.mediebedriftene.no.

Menon, Sreelatha, 'Ear to the ground', *Business Standard*, 4 February 2007.

Monbiot, George, 'The day the world went mad', *Guardian*, 29 August 2012, available at www.guardian.co.uk/environment/georgemonbiot/2012/aug/29/day-world-went-mad.

Morton, Thomas A., et al., 'The future that may (or may not) come: how framing changes responses to uncertainty in climate change communications', *Global Environmental Change* 21 (2011), pp. 103–9.

Moser, Susanne, 'Communicating climate change: history, challenges, progress and future directions', *Wiley Interdisciplinary Reviews: Climate Change*, Volume 1, January/February 2010.

Moser, Susanne and Lisa Dilling, 'Making climate hot: communicating the urgency and challenge of global climate change', *Environment* 46 (2004), pp. 32–46.

Moser, Susanne and Lisa Dilling (eds), *Creating a Climate for Change: Communicating Climate Change and Facilitating Social Change*, Cambridge: Cambridge University Press, 2007.

MRUC and Hansa Research, Indian Readership Survey, 2012, 'IRS 2012 Q3 Topline Findings', available at http://mruc.net/irs2012q3_topline_findings.pdf.

National Intelligence Council, *Global Trends 2030: Alternative Worlds*, Washington, December 2012.

Navroz, Dubash, 'The politics of climate change in India: Narratives of equity and co-benefits', *WIREs Climate Change* 4 (2013), pp. 191–201.

NCTJ, 'Journalists at work', London, 2013, available at NCTJ website.

Nerlich, Brigitte, 'The impact of earthquakes on making science public', 2012, available at http://blogs.nottingham.ac.uk/makingsciencepublic/2012/10/25/the-impact-of-earthquakes-on-making-science-public.

Newman, Nic, 'Reuters Institute Digital News Report 2012: Tracking the Future of News', RISJ, 2012, available at https://reutersinstitute.politics.ox.ac.uk/fileadmin/documents/Publications/Other_publications/Reuters_Institute_Digital_Report.pdf.

Nisbet, Matthew C., 'Communicating climate change: Why frames matter for public engagement', *Environment Magazine* 51 (2) (2009), p. 12.

Nisbet, Matthew C. and Chris Mooney, 'Framing science', *Science* 316 (2007), p. 56.

Olausson, Ulrika, 'Global warming – global responsibility? Media frames of collective action and scientific uncertainty', *Public Understanding of Science*, 18 (2009), p. 421.

O'Neill, Saffron and Sophie Nicholson-Cole, '"Fear won't do it": promoting positive engagement with climate change through visual and iconic representations', *Science Communication* 30 (2009), pp. 355–79.

Oreskes, Naomi and Erik M. Conway, *The Merchants of Doubt*, New York: Bloomsbury, 2010.

Otto, Friederike E.L., et al, 'Reconciling two approaches to attribution of the 2010 Russian heat wave', *Geophysical Research Letters* 39 (L04702) (2012).

Painter, James, *Summoned by Science: Reporting Climate Change at Copenhagen and Beyond*, Oxford: RISJ, 2010.

Painter, James, *Poles Apart: The International Reporting of Climate Scepticism*, Oxford: RISJ, 2011.

Painter, James and Teresa Ashe, 'Cross-national comparison of the presence of climate scepticism in the print media in six countries, 2007–10', *Environmental Research Letters* 7 (2012), 04400.

Painter, James and Neil Gavin, *Climate Scepticism in the British Press*, unpublished manuscript, 2013.

Pall, Pardeep, et al, 'Anthropogenic greenhouse gas contribution to flood risk in England and Wales in autumn 2000', *Nature* 470 (2011), pp. 382–5.

Palmer, Lisa, 'Risky business: Climate concerns in boardroom, but in business magazines?', *Yale Climate Forum*, 12 October 2012.

Pearce, Fred, 'On climate models, the case for living with uncertainty', *Environment* 360 (5 October 2010).

Petersen, Arthur, 'Being an expert in the age of uncertainty', 3 August 2012, available at http://blogs.lse.ac.uk/politicsandpolicy/archives/25741.

Pew Research Center for the People and the Press, 'Fewer Americans see solid evidence of global warming', 2009, available at http://peoplepress.org/report/556/global-warming.

Pidcock, Roz, 'Can the public make sense of uncertainty in weather and climate prediction?', *Carbon Brief*, 11 October 2012a.

Pidcock, Roz, 'How well have the media covered hurricane Sandy?', *Carbon Brief*, 6 November 2012b.

Pidgeon, Nick and Baruch Fischhoff, 'The role of social and decision sciences in communicating uncertain climate risks', *Nature Climate Change* 1 (2011), pp. 35–41.

Pielke, Roger A., Jr.,*The Honest Broker: Making Sense of Science in Policy and Politics*, Cambridge: Cambridge University Press, 2007.

Pollack, Henry, *Uncertain Science... Uncertain World*, Cambridge: Cambridge University Press, 2003.

Poortinga, Wouter, et al, 'Uncertain climate: An investigation into public scepticism about anthropogenic climate change', *Global Environmental Change* 21 (3) (2011), pp. 1015–24.

Porritt, Jonathon, 'Why do we play down the horror of climate change?', *Guardian*, 11 October 2012, available at www.guardian.co.uk/environment/2012/oct/11/why-play-down-horror-climate-change.

Pugliese, Anita and Julie Ray, 'Top-Emitting Countries Differ on Climate Change Threat', *Gallup World*, 7 December 2009.

Pugliese, Anita and Julie Ray, 'Fewer Americans, Europeans view global warming as a threat', *Gallup World*, 20 April 2011.

Radford, Tim, 'US: the climate IS warming', *Climate News Network*, 17 January 2013.

Raupach, Michael, et al, 'Climate change poised to feed on itself', *Sydney Morning Herald*, 1 August 2009.

Rees, Daniel, 'TNS Gallups Klimabarometer 2011', 2012.

Reusswig, Fritz and Lutz Meyer-Ohlendorf, 'Adapting to what? Climate change impacts on Indian megacities and the local Indian climate change discourse', in William G. Holt (ed.) *Urban Areas and Global Climate Change (Research in Urban Sociology, Volume 12)*, Emerald Group Publishing Limited, 2012, pp. 197–219.

Revkin, Andrew, 'The daily planet: Why the media stumble over the environment', in Deborah Blum et al (eds), *A Field Guide for Science Writers*, Oxford: Oxford University Press, 2005.

Revkin, Andy, 'A new middle stance emerges in debate over climate', *New York Times*, 1 January 2007.

Revkin, Andy, 'Planet-scale risk and the "Steve Schneider Memorial Exercise"', *Dot Earth Blog*, 4 August 2010.

Riesch, Hauke and David Spiegelhalter, 'Careless pork costs lives': risk stories from science to press release to media', *Health, Risk & Society* 13 (1) (February 2011), pp. 47–64.

Rose, David, 'The great green con no 1', *Mail on Sunday*, 16 March 2013.

Rosenthal, Elisabeth and Andrew C. Revkin, 'Science panel says global warming is "unequivocal"', *New York Times*, 3 February 2007.

Roy, Prannoy, RISJ lecture, Oxford, November 2012.

Rudd, Kevin, 'Climate change: the great moral challenge of our generation', National Library of Australia, 2007.

Ryghaug, Marianne, 'Some like it hot – Konstruksjon av kunnskap om klimaendringer i norske aviser', *Norsk medietidsskrift* 13 (2006), pp. 197–214.

Ryghaug, Marianne and Tomas Moe Skjølsvold, 'Achieving a balanced view? A comparative analysis of newspaper coverage of climate change in Norway and Sweden, 2002–2008', IOP Conference Series Earth and Environment, 2009.

Sandman, Peter M. and Jody Lanard, 'Explaining and proclaiming uncertainty: risk communication lessons from Germany's deadly E. coli outbreak', 2011, available at www.psandman.com/col/GermanEcoli.htm.

Schiermeier, Quirin, 'The real holes in climate science', *Nature* 463 (2010), pp. 284–7.

Schneider, Stephen, 'Science as a contact sport', *National Geographic*, Washington, 2009, pp. 148–54.

Science Media Centre, '10 best practice guidelines for reporting science and health stories', London, 2012, available at www.sciencemediacentre.org.

Sethi, Nithin, 'In 18 years, 505 less water for you', *The Times of India*, 5 April 2007.

Shuckburgh, Emily, Rosie Robison and Nick Pidgeon, *Climate Science, the Public and the News Media*, Swindon: Living With Environmental Change, 2012.

Shukman, David, 'First report on UK climate impact', BBC website, 26 January 2012.

Siegel, Matt, 'Report blames climate change for extremes in Australia', *New York Times*, 4 March 2013.

Somerville, Richard C.J. and Susan Joy Hassol, 'Communicating the science of climate change', *Physics Today* 64 (10) (2011), pp. 48–53.

Spence, Alexa, et al, 'Public perceptions of climate change and energy futures in Britain: Summary findings of a survey conducted in January–March 2010', Technical Report (Understanding Risk Working Paper 10-01), Cardiff: School of Psychology, 2010.

Spiegelhalter, David and Hauke Riesch, 'Don't know, can't know: embracing deeper uncertainties when analysing risks', *Philosophical Transactions of the Royal Society A* 369 (2011), pp. 4730–50.

Spiegelhalter, David, et al, 'Visualizing uncertainty about the future', *Science* 333 (9 September 2011), pp. 1393–400.

Stainforth, David, 'Climate science in the spotlight may not be such a bad thing', *Guardian* website, 12 February 2010.

Statistics Norway, 'Emissions of greenhouse gases, 1990–2011, final figures', 2013, available at www.ssb.no/en/natur-og-miljo/statistikker/klimagassn/aar-endelige.

Steffen, Will, *The Angry Summer*, Climate Commission, 2013.

Stephens, Elisabeth, et al, 'Communicating probabilistic information from climate model ensembles lessons from numerical weather prediction', *WIREs Climate Change* 3 (2012), pp. 409–26.

Stott, Peter A., et al, 'Human contribution to the European heatwave of 2003', *Nature* 432 (2004), pp. 610–14.

Sydney Morning Herald, 'Australia ratifies Kyoto Protocol', *Sydney Morning Herald*, 3 December 2007.

Sydney Morning Herald, 'Gillard rules out imposing carbon tax', *Sydney Morning Herald*, 17 August 2010.

The Times of India, 'No climate sceptics in Lok Sabha', *The Times of India*, 4 February 2009, available at http://articles.timesofindia.indiatimes.com/2009-12-04/india/28061875_1_climate-change-mitigation-and-adaptation-global-warming.

The Times of India, 'Mumbai, Miami on list for big weather disasters', *The Times of India*, 29 March 2012.

Trenberth, Kevin, 'More knowledge, less certainty', *Nature Reports* 4 (February 2010), pp. 20–1.

Truc, Olivier, 'Les pays arctiques sur la voie de la coopération', *Le Monde*, 13 May 2011.

Vergano, Dan and Patrick O'Driscoll, 'Is Earth near its "tipping points" from global warming?', *USA Today*, 3 April 2007.

Ward, Robert E.T., 'Good and bad practice in the communication of uncertainties associated with the relationship between climate change and weather-related natural disasters', *Geological Society* 305 (2008), pp. 19–37.

Warren, Matthew, 'Doom merchants', *Australian*, 3 February 2007.

Weintrobe, Sally (ed.) *Engaging with Climate Change*, London: Routledge, 2013.

Whitmarsh, Lorraine, 'Scepticism and uncertainty about climate change', *Global Environmental Change* 21 (2) (May 2011), pp. 690–700.